Springer Theses

Recognizing Outstanding Ph.D. Research

Aims and Scope

The series "Springer Theses" brings together a selection of the very best Ph.D. theses from around the world and across the physical sciences. Nominated and endorsed by two recognized specialists, each published volume has been selected for its scientific excellence and the high impact of its contents for the pertinent field of research. For greater accessibility to non-specialists, the published versions include an extended introduction, as well as a foreword by the student's supervisor explaining the special relevance of the work for the field. As a whole, the series will provide a valuable resource both for newcomers to the research fields described, and for other scientists seeking detailed background information on special questions. Finally, it provides an accredited documentation of the valuable contributions made by today's younger generation of scientists.

Theses are accepted into the series by invited nomination only and must fulfill all of the following criteria

- They must be written in good English.
- The topic should fall within the confines of Chemistry, Physics, Earth Sciences, Engineering and related interdisciplinary fields such as Materials, Nanoscience, Chemical Engineering, Complex Systems and Biophysics.
- The work reported in the thesis must represent a significant scientific advance.
- If the thesis includes previously published material, permission to reproduce this must be gained from the respective copyright holder.
- They must have been examined and passed during the 12 months prior to nomination.
- Each thesis should include a foreword by the supervisor outlining the significance of its content.
- The theses should have a clearly defined structure including an introduction accessible to scientists not expert in that particular field.

Thomas Scheby Kuhlman

The Non-Ergodic Nature of Internal Conversion

An Experimental and Theoretical Approach

Doctoral Thesis accepted by
the Technical University of Denmark

 Springer

Author
Dr. Thomas Scheby Kuhlman
Department of Chemistry
Technical University of Denmark
Kgs. Lyngby
Denmark

Supervisors
Prof. Dr. Klaus B. Møller
Department of Chemistry
Technical University of Denmark
Kgs. Lyngby
Denmark

Prof. Dr. Theis I. Sølling
Department of Chemistry
University of Copenhagen
Copenhagen
Denmark

ISSN 2190-5053 ISSN 2190-5061 (electronic)
ISBN 978-3-319-03386-0 ISBN 978-3-319-00386-3 (eBook)
DOI 10.1007/978-3-319-00386-3
Springer Cham Heidelberg New York Dordrecht London

Printed on acid-free paper

Springer is part of Springer Science+Business Media (www.springer.com)

Parts of this thesis have been published in the following journal articles:

I *Coherent Motion Reveals Non-Ergodic Nature of Internal Conversion between Excited States*
T. S. Kuhlman, T. I. Sølling, K. B. Møller, *ChemPhysChem* **13**, 820–827 (2012)

II *Between Ethylene and Polyenes - The Non-Adiabatic Dynamics of cis-Dienes*
T. S. Kuhlman, W. J. Glover, T. Mori, K. B. Møller, T. J. Martínez, *Faraday Discuss.* **157**, 193–212 (2012)

III *Symmetry, Vibrational Energy Redistribution and Vibronic Coupling: The Internal Conversion Processes of Cycloketones*
T. S. Kuhlman, S. P. A. Sauer, T. I. Sølling, K. B. Møller, *J. Chem. Phys.* **137**, 22A522 (2012)

IV *Surprising Intrinsic Photostability of the Disulfide Bridge Common in Proteins*
A. B. Stephansen, R. Y. Brogaard, T. S. Kuhlman, L. B. Klein, J. B. Christensen, T. I. Sølling, *J. Am. Chem. Soc.* **134**, 20279–20281 (2012)

V *Pulling the Levers of Photophysics: How Structure Controls the Rate of Energy Dissipation*
T. S. Kuhlman, M. Pittelkow, T. I. Sølling, K. B. Møller, *Angew. Chem. Int. Ed.* **52**, 2247–2250 (2013)

Parts of this thesis are discussed in the following manuscripts in preparation:

VI *Applying the Full Multiple Spawning Code to Absorption Spectroscopy*
T. S. Kuhlman, A. Witt, C. Evenhuis, K. B. Møller, T. J. Martínez, *to be submitted to J. Chem. Phys.* (2013)

VII *Simulating the Time-Resolved Photoelectron Spectrum of Cyclobutanone*
T. S. Kuhlman, T. I. Sølling, K. B. Møller, *to be submitted to Chem. Phys. Lett.* (2013)

VIII *Photoinduced Dynamics of Cyclopentadiene: The Role of Substitution at the sp3-Position*
T. J. A. Wolf, T. S. Kuhlman, O. Schalk, A.-N. Unterreiner, T. J. Martínez, K. B. Møller, A. Stolow, *in preparation* (2013)

Supervisors' Foreword

The deactivation of electronically excited states is an extremely important process in all branches of science that involve the interaction between light and matter. Whether it is in solar cells or when DNA is exposed to light, the mechanism by which the systems disperse the energy is pivotal for the subsequent faith of the involved molecules. If the initially excited electronic state stays energized for an extended period of time, new and effective pathways for decomposition open up. With femtosecond resolution it is possible to study the fastest of events in chemistry—such as for example the transition from one electronically excited state to another, which is the focus of this thesis. Using both experimental and theoretical methods Thomas S. Kuhlman has nailed down that the electronic transition proceeds before the entire set of available degrees of freedom are active. It is as simple as that! With an outstanding new treatment of our experimental data it has even been possible to map out which degrees of freedom are in play during the electronic transition and thus devise a recipe to speed up or slow down an electronic transition. There is no doubt that this finding will extend far into a broad range of chemical disciplines that deal with light and we are therefore happy that the thesis will reach a much broader audience though the Springer award. We can reveal that the idea has already given rise to a very promising study that explains why the tertiary structure of proteins has survived evolution; the reason is simply that a specific and thus highly non-statistical decay pathway exists. Before we give away the ending of the story we cannot encourage you enough to read about all of the findings in this truly outstanding thesis.

Lyngby, March 2013
Copenhagen
Klaus B. Møller
Theis I. Sølling

Preface

This thesis has been submitted to the Department of Chemistry, Technical University of Denmark, in partial fulfillment of the requirements for the Ph.D. degree in the subject of Chemistry. The work presented herein was carried out at the Department of Chemistry, Technical University of Denmark, and the Department of Chemistry, University of Copenhagen, over the period from January 2010 to December 2012 under the joint supervision of Prof. Klaus B. Møller and Prof. Theis I. Sølling. In addition, work was carried out in the group of Prof. Todd J. Martínez, Department of Chemistry and PULSE Institute, Stanford University, and SLAC National Accelerator Laboratory, during a seven-month stay in 2011.

Acknowledgments

As is the case with many things in life, my way into the world of ultrafast chemical dynamics was not the result of a well thought out plan but was due to a combination of chance and opportunity. I am very grateful to have ended up in a fruitful academic environment under the supervision of two very different supervisors, Klaus B. Møller and Theis I. Sølling. Klaus and Theis in many ways complement each other as supervisors and scientists which has allowed me to take on combined theoretical and experimental endeavors during my Ph.D. studies. I am grateful to Klaus and Theis for the many stimulating discussions and for the opportunities and support they have provided during the last three years.

During my Ph.D. studies, I was fortunate to be able to visit the group of Prof. Todd J. Martínez at Stanford University. The large research group and the strong academic environment at Stanford University provided the basis for a great scientific experience. I am in particular grateful to Alexander Witt, Christian Evenhuis, Toshifumi Mori, and Will Glover for fruitful collaborations and to the rest of the group, in particular Christine Isborn, for making my time in California very enjoyable.

At the University of Copenhagen, I have had the pleasure to share the office B416 with many other students as well as an occasional post.doc. In particular, I would like to give thanks to Christer Bisgaard who very patiently showed me the way around the laser laboratory and always took his time to answer my questions. I am grateful to Rasmus Brogaard, Martin Rosenberg, and Ditte Linde for contributing to the thriving scientific and social environment in B416.

Asmus Dohn and Jakob Petersen as well as Chuan-Cun Shu and Ulf Lorenz, with whom I shared an office at the Technical University of Denmark, are thanked for the many good times at DTU.

Over the course of the last three years, I have had the pleasure to collaborate with several people who have generously assisted me with their knowledge and skills. I would like to thank Stephan P. A. Sauer and Michael Pittelkow, University of Copenhagen, for collaboration on the computational and experimental aspects of the cycloketone project, respectively, Oliver Schalk, Ludwig Maximilian University of Munich, and Thomas Wolf, Karlsruhe Institute of Technology, for collaboration on the cyclopentadiene project, and Senior Lecturer Graham Worth, University of Birmingham, for providing a developmental version of the Heidelberg

MCTDH code. Prof. Jure Demsar and Michael Koerber, University of Konstanz, and Verner Thorsmølle, Rutgers University, are thanked for the continuing collaboration on the pentacene project which is not presented in this thesis but has nevertheless amassed some of my time and attention during the last three years.

Finally, I am eternally grateful to my girlfriend Anja Kopalska for her tireless support in particular during the last part of the studies.

Copenhagen, December 2012   Thomas Scheby Kuhlman

Contents

Abbreviations

2-MeCB	2-Methylcyclobutanone
2-MeCP	2-Methylcyclopentanone
3-EtCP	3-Ethylcyclopentanone
3-MeCP	3-Methylcyclopentanone
AIMS	Ab Initio Multiple Spawning
au	Atomic Unit(s)
BBO	Beta Barium Borate
CASPT2	Complete Active Space Second Order Perturbation Theory
CASSCF	Complete Active Space Self-Consistent Field
CB	Cyclobutanone
CC2	Coupled-Cluster Singles and Approximate Doubles
CC3	Coupled-Cluster Singles, Doubles and Approximate Triples
CCS	Coupled-Cluster Singles
CCSD	Coupled-Cluster Singles and Doubles
CCSDR(3)	Coupled-Cluster Singles, Doubles and Approximate (noniterative) Triples
CH	Cyclohexanone
CM	Classical Mechanics
CP	Cyclopentanone
CPD	Cyclopentadiene
CW	Continuous Wave
DC	Direct Current
DD-vMCG	Direct Dynamics Variational Multi-Configuration Gaussian Wavepacket
DMIPA	N,N-Dimethylisopropylamine
DVR	Discrete Variable Representation
EOM	Equation of Motion
EOM-CCSD	Equation of Motion Coupled-Cluster Singles and Doubles
EOMIP	Equation of Motion Ionization Potential
FC	Franck–Condon

FMS	Full Multiple Spawning
FWHM	Full Width at Half Maximum
G-MCTDH	Gaussian Multi-Configuration Time-Dependent Hartree
IFG	Independent First Generation
IP	Ionization Potential
IVR	Internal Vibrational Energy Redistribution
MCP	Micro-Channel Plate
MCSCF	Multi-Configuration Self-Consistent Field
MCTDH	Multi-Configuration Time-Dependent Hartree
Me_4CPD	1,2,3,4-Tetramethylcyclopentadiene
Me_6CPD	Hexamethylcyclopentadiene
MECI	Minimum Energy Conical Intersection
MS-MR-CASPT2	Multi-State Multi-Reference Complete Active Space Second Order Perturbation Theory
Nd:YLF	Neodymium-doped Yttrium Lithium Fluoride ($LiYF_4$)
$Nd:YVO_4$	Neodymium-doped Yttrium Orthovanadate (YVO_4)
NEB	Nudged Elastic Band
QM	Quantum Mechanics
SA-Ne-CAS(n,m)SCF	State-Averaged Complete Active Space Self-Consistent Field for which the orbitals are optimized simultaneously for Ne states using an active space of n electrons in m orbitals
SFG	Sum-Frequency Generation
SHG	Second-Harmonic Generation
TBF	Trajectory Basis Function
Ti:Sapphire	Titanium-doped Sapphire (Al_2O_3)
TOF	Time-of-Flight
TOPAS	Collinear (Travelling Wave) Optical Parametric Amplifier of White-Light (Super) Continuum
TRMS	Time-Resolved Mass Spectrometry
TRPES	Time-Resolved Photoelectron Spectroscopy
UV	Ultraviolet
VCHAM	Vibronic Coupling Hamiltonian
VCHFIT	Program for fitting the parameters of the vibronic coupling Hamiltonian to adiabatic potential energy surfaces
vMCG	Variational Multi-Configuration Gaussian Wavepacket
XC	Cross-Correlation

Part I
Introductory Remarks

Chapter 1
Introduction

This thesis is concerned with the ultrafast dynamics of coupled electronic and nuclear vibrational motion that unfold when molecules are excited by light. Herein, ultrafast denote processes taking place on the femtosecond to picosecond timescale—the fundamental timescale of nuclear vibrational motion—ubiquitous in chemical reaction dynamics [1].

1.1 Ultrafast Excited-State Dynamics

Ultrafast photoinduced processes result from the complex charge and energy flows which are central to chemical reaction dynamics [2, 3]. Dissociation was among the first of such processes to be followed in real time in the seminal works on ICN and NaI from the group of Nobel laureate Ahmed Zewail [4–7]. The interplay between charge and energy can lead to many other processes such as isomerization, which in retinal initiates the process of vision, [8–10] bond formation, which in stacked DNA bases causes mutagenic photolesions, [11–13] or internal conversion, by which electronic energy can be dissipated in e.g. conjugated molecules [14–17].

One distinguishing feature of these ultrafast excited-state processes is that they involve more than one potential energy surface—a concept we will introduce in more detail in the next chapter. By this merit, the work presented inhere is not only relevant for reactions initiated by short laser pulses but indeed for a large range of chemical and physical processes [18]. As an example take electron transfer, where, following the Franck-Condon principle, the transfer of charge takes place near the intersection of the potential energy surface of the reactant and that of the product, cf. Fig. 1.1a [19–21]. Similarly, atomic and molecular reactive scattering [22–24] and reactive scattering off surfaces, [25–27] which is of central importance in heterogeneous catalysis, [28, 29] often fall in this range, cf. Fig. 1.1b. Both electron transfer and reactive scattering are fundamental to the very notion of chemistry as are the energy transfer processes on which we will focus.

T. S. Kuhlman, *The Non-Ergodic Nature of Internal Conversion*, Springer Theses, DOI: 10.1007/978-3-319-00386-3_1, © Springer International Publishing Switzerland 2013

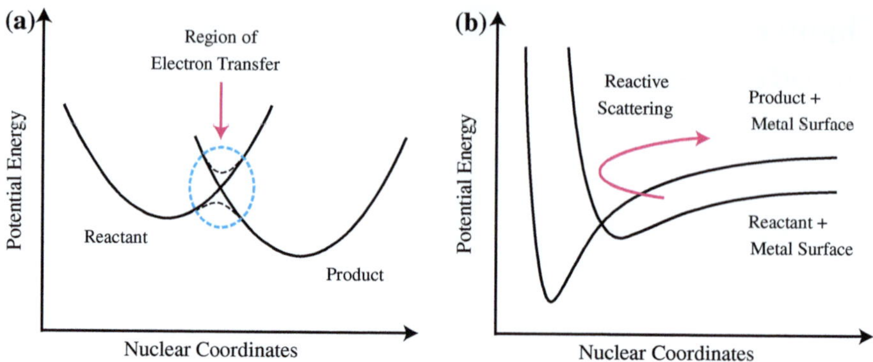

Fig. 1.1 Examples of processes involving more than one potential energy surface. (**a**) Electron transfer takes place near the intersection of two potential energy surfaces as the system changes from the potential energy surface of the reactant to that of the product [19], and (**b**) reactive scattering off a metal surface [28, p. 421]

Another distinguishing feature of ultrafast processes is the very timescale on which they take place. A relevant question in this context is whether the short timescale directly reflects a restriction in the scope of the dynamics that unfold, or more precisely, if the short timescale is an indication that only a restricted region of phase space is explored before the process takes place? Reduced-space dynamics is a reflection of ergodicity breaking which will be a central theme in this thesis. The ergodic hypothesis of thermodynamics states that a system left to itself will sooner or later pass through every energetically available point in phase space [30, 31, pp. 100–101, pp. 545–555]. A system is termed ergodic if it satisfies this hypothesis. Ergodicity breaking occurs if the timescale is too short for all these points to be visited – i.e. ergodicity is essentially a question of relative timescales. Herein, we are concerned with processes initiated by a femtosecond laser pulse. Such activation creates a nuclear wavepacket, or equivalently a phase space distribution, which is localized in space and energy due to the coherence of the short laser pulse. If the timescale for a subsequent process is shorter than the timescale for loss of coherence of the wavepacket, or equivalently for randomization of the initial phase space distribution, the process is non-ergodic, and the dynamics will be continuously localized, cf. Fig. 1.2(a) [32, 33]. The (non-)ergodicity of a process thus rests on the relative timescales of internal vibrational energy redistribution (IVR) and that of the process in question. If the timescale for IVR is relatively fast, energy will be redistributed efficiently and uniformly before and during the process whereby a microcanonical ensemble will be maintained, and the process will exhibit ergodic behavior, cf. Fig. 1.2(b). In contrast, the observation of localized coherent dynamics following preparation by a short laser pulse entails a given process to be fast on a timescale relative to that of IVR and thus to exhibit non-ergodic behavior.

The intent of this thesis is to exhibit the non-ergodic nature of ultrafast excited-state dynamics. In particular, we are interested in the process of internal conversion where a molecule passes from one potential energy surface to another such that

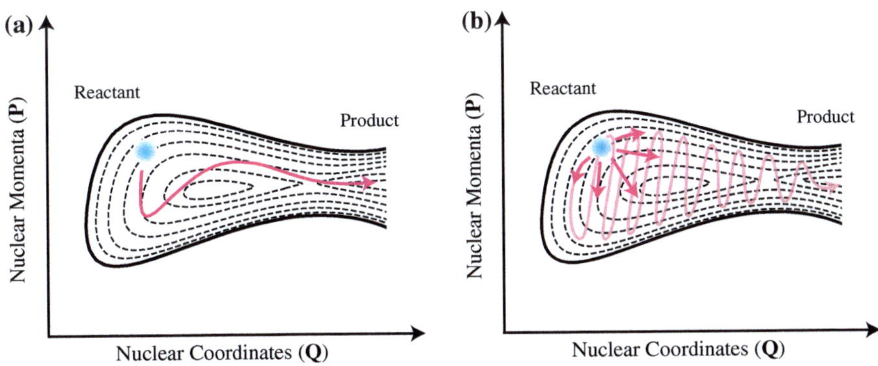

Fig. 1.2 Illustration of non-ergodic and ergodic dynamics in phase space following initial preparation of a localized phase space distribution by a femtosecond laser pulse. (**a**) The initial distribution stays localized as the reaction proceeds, and (**b**) fast redistribution of energy to the entire phase space of the reactant ensures a microcanonical ensemble is maintained during reaction

electronic energy is transformed into nuclear kinetic energy. With the concept of non-ergodicity at hand, we seek to understand the process and its timescale from simple structural and energetic properties. Given a molecular structure, can we determine which degrees of freedom are important in the internal conversion process, i.e. where are the dynamics localized? In this respect, how does a particular substitution affect the rate of internal conversion? In effect, can we propose rules of thumb as to how structural and energetic parameters affect the internal conversion process?

To be able to address these questions, we carry out simulations and experiments on a series of molecules. As an example of weak coupling between the potential energy surfaces involved, we investigate the $S_2 \rightarrow S_1$ transition between a Rydberg and a valence state in seven cycloketones which differ by ring size as well as substitution pattern. As examples of strong coupling, we investigate the $S_1 \rightarrow S_0$ transition in three cyclopentadienes, which indirectly also involves the S_2 state, as well as the $S_1 \rightarrow S_0$ transition in dithiane. These systems touch upon not only electronic energy dissipation but also isomerization and bond dissociation.

The thesis is organized as follows: as a starting point, we will in Sect. 1.2 introduce the Born-Oppenheimer approximation which provides us with the concept of a potential energy surface essential to our mechanistic trajectory-based picture of molecular dynamics. In Sect. 1.3, we will detail whereby the methods of time-resolved mass spectrometry and photoelectron spectroscopy can be used to investigate the dynamics of coupled electronic and nuclear vibrational motion, and in Chap. 2, we describe the experimental setup used to perform these experiments. Methods of data analysis are presented in Chap. 3. In Chap. 4, we provide an introduction to the Multi-Configuration Time-Dependent Hartree and Ab Initio Multiple Spawning methodologies for simulating molecular dynamics, and in Chap. 5, it is detailed how these simulations can be used to calculate time-resolved photoelectron spectra. The part of the thesis dealing with the theoretical and computational approaches is completed by Chap. 6, in which we provide a short introduction to the methods of electronic structure calculation employed. If the reader is familiar with the experimental and/or theoretical methods, the relevant chapters can be skipped. The results and discussion part of the thesis is divided into three main chapters detailing the findings relating to the cycloketones, Chap. 7, the cyclopentadienes, Chap. 8, and dithiane, Chap. 9. Finally, a summarizing discussion of the findings is presented in Chap. 10.

We will throughout this thesis employ atomic units in formulas, i.e. the reduced Planck's constant $\hbar = 1$, the elementary charge $e = 1$, and the electron mass $m_e = 1$.

1.2 The Born-Oppenheimer Approximation

To describe the time-evolution of a molecular system, we need to solve the time-dependent Schrödinger equation for the total electronic and nuclear wavefunction

$$i\frac{\partial}{\partial t}\Psi(\mathbf{r}, \mathbf{R}, t) = \hat{H}\Psi(\mathbf{r}, \mathbf{R}, t) \tag{1.1}$$

with electronic and nuclear coordinates denoted by $\mathbf{r}$ and $\mathbf{R}$ respectively. Throughout this thesis, we will consider the non-relativistic molecular Hamiltonian for both nuclei and electrons given by

$$\hat{H} = \hat{T}_n + \hat{T}_{el} + \hat{V} = \hat{T}_n + \hat{H}_{el} \tag{1.2}$$

$\hat{T}_n$ and $\hat{T}_{el}$ are the nuclear and electronic kinetic energy operators respectively, and $\hat{V}$ is the potential energy operator for both nuclei and electrons. In the Born representation, the wavefunction is expanded according to [34–36]

$$\Psi(\mathbf{r}, \mathbf{R}, t) = \sum_v \Phi^{(v)}(\mathbf{R}, t)\psi^{(v)}(\mathbf{r}; \mathbf{R}) \tag{1.3}$$

The electronic functions $\psi^{(v)}(\mathbf{r}; \mathbf{R})$ depend parametrically on the nuclear coordinates $\mathbf{R}$. For a given value of $\mathbf{R}$, these functions are solutions to the clamped-nuclei eigenvalue equation

$$\hat{H}_{el}\psi^{(v)}(\mathbf{r}; \mathbf{R}) = V^{(v)}(\mathbf{R})\psi^{(v)}(\mathbf{r}; \mathbf{R}) \tag{1.4}$$

The functions $\psi^{(v)}(\mathbf{r}; \mathbf{R})$ constitute a complete set of electronic functions and define the electronic energies $V^{(v)}$ from the above equation. Inserting the representation of the total wavefunction in Eq. 1.3 into the time-dependent Schrödinger equation yields an equation for the nuclear functions

$$i\frac{\partial}{\partial t}\Phi^{(v)}(\mathbf{R}, t) = \left(\hat{T}_n + V^{(v)}(\mathbf{R})\right)\Phi^{(v)}(\mathbf{R}, t) - \sum_w \hat{\Lambda}^{(v,w)}\Phi^{(w)}(\mathbf{R}, t) \tag{1.5}$$

Here, $\hat{\Lambda}^{(v,w)}$ are derivative coupling operators. If one makes an adiabatic approximation and replaces the Born representation of the wavefunction by a single term, i.e.

$$\Psi(\mathbf{r}, \mathbf{R}, t) = \Phi^{(v)}(\mathbf{R}, t)\psi^{(v)}(\mathbf{r}; \mathbf{R}) \tag{1.6}$$

the equation for the nuclear functions reduces to

$$i\frac{\partial}{\partial t}\Phi^{(v)}(\mathbf{R}, t) = \left(\hat{T}_n + V^{(v)}(\mathbf{R}) - \hat{\Lambda}^{(v,v)}\right)\Phi^{(v)}(\mathbf{R}, t) \tag{1.7}$$

Together with Eq. 1.6 for the wavefunction, this equation constitutes the Born-Huang adiabatic approximation [37]. If the on-diagonal derivative coupling term is furthermore neglected, one arrives at the Born-Oppenheimer adiabatic approximation [38]

$$i\frac{\partial}{\partial t}\Phi^{(v)}(\mathbf{R}, t) = \left(\hat{T}_n + V^{(v)}(\mathbf{R})\right)\Phi^{(v)}(\mathbf{R}, t) \tag{1.8}$$

The appropriateness of the separation of nuclear and electronic motion in the Born-Oppenheimer approximation is a consequence of the large difference between nuclear and electronic masses which usually results in the neglected derivative coupling terms being very small. However, this is not always the case as we will see below. The solutions to Eq. 1.4 provide an N-dimensional hypersurface of the electronic energy for a given electronic state v where N refers to the number of internal nuclear degrees of freedom. In the Born-Oppenheimer approximation, this hypersurface acts as a potential energy surface on which the nuclei move, and it is thus a central concept for the understanding of molecular dynamics from a mechanistic point of view.

In rectilinear coordinates, the derivative or non-adiabatic coupling operators of Eq. 1.5 for which $v \neq w$ can be given in terms of first- and second-order couplings

$$\hat{\Lambda}^{(v,w)} = 2\hat{D}^{(v,w)} + \hat{G}^{(v,w)} \tag{1.9}$$

with

$$\hat{D}^{(v,w)} = \sum_{\kappa}^{N} \frac{1}{2m_\kappa}\hat{d}_\kappa^{(v,w)}\frac{\partial}{\partial R_\kappa} \tag{1.10}$$

$$\hat{G}^{(v,w)} = \sum_{\kappa}^{N} \frac{1}{2m_\kappa}\left\langle\psi^{(v)}\left|\frac{\partial^2}{\partial R_\kappa^2}\right|\psi^{(w)}\right\rangle \tag{1.11}$$

Corrections to the single-surface adiabatic approximation of Eq. 1.6 can be introduced by evaluating these mass-dependent operators which couple the adiabatic electronic states giving rise to non-adiabatic nuclear dynamics. The derivative coupling vector $\mathbf{d}^{(v,w)}$, which collects the elements $\hat{d}_\kappa^{(v,w)}$ from Eq. 1.10, is given by [36, 39]

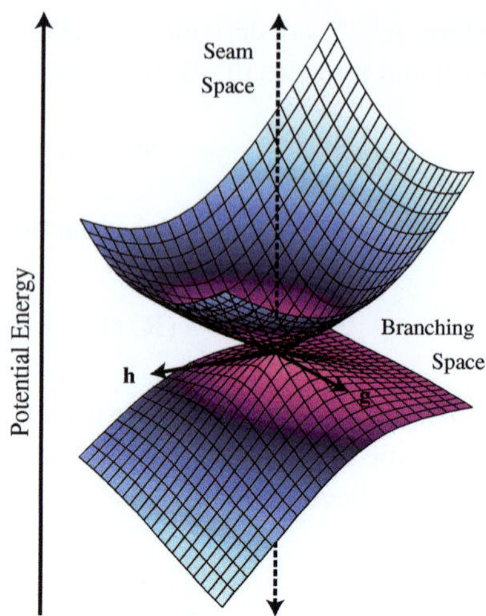

Fig. 1.3 Example of a peaked conical intersection in the branching space of the gradient difference and scaled derivative coupling vectors. These coordinates linearly lift the degeneracy of the two potential energy surfaces. The degeneracy is maintained in the seam or intersection space of nuclear coordinates, which is orthogonal to the branching space, represented by the dashed line [44, 40]

$$\mathbf{d}^{(v,w)}(\mathbf{R}) = \langle \psi^{(v)} | \nabla_n \psi^{(w)} \rangle = \frac{\langle \psi^{(v)} | (\nabla_n \hat{H}_{el}) | \psi^{(w)} \rangle}{V^{(w)}(\mathbf{R}) - V^{(v)}(\mathbf{R})} \tag{1.12}$$

where ∇_n denotes the nuclear derivative operator. The last form of the derivative coupling vector exhibits that the non-adiabatic coupling is negligible for well-separated potential energy surfaces. However, the coupling diverges when the surfaces come into close proximity and is singular at conical intersections where two or more surfaces become degenerate [40–42]. Figure 1.3 depicts a conical intersection in the so-called branching space of the scaled derivative coupling and gradient difference vectors given by

$$\mathbf{h}^{(v,w)}(\mathbf{R}) = \langle \psi^{(v)} | (\nabla_n \hat{H}_{el}) | \psi^{(w)} \rangle \tag{1.13}$$

$$\mathbf{g}^{(v,w)}(\mathbf{R}) = \nabla_n \left(V^{(w)}(\mathbf{R}) - V^{(v)}(\mathbf{R}) \right) \tag{1.14}$$

From this two-dimensional representation, it is not evident that conical intersections are indeed extended high-dimensional seams of dimension $N - 2$ in nuclear coordinate space. These seams act as effective doorways connecting the adiabatic potential energy surfaces. Thus, the presence of a conical intersection indicates the possibility for ultrafast non-adiabatic dynamics—nuclear dynamics involving more than one adiabatic surface—due to the diverging coupling between the adiabatic electronic states in their vicinity [41, 43].

Whereas the solutions to Eq. 1.4 for different v's provide a set of adiabatic electronic states, a diabatic electronic basis can be defined by a nuclear coordinate dependent unitary transformation of the adiabatic wavefunctions

$$\widetilde{\psi}^{(v)}(\mathbf{r}; \mathbf{R}) = \sum_w U^{(v,w)}(\mathbf{R}) \psi^{(w)}(\mathbf{r}; \mathbf{R}) \tag{1.15}$$

under the requirement that $\mathbf{d}^{(v,w)} = 0$ in the new basis. For triatomic or larger molecules, this is generally not possible [45]. Instead, one can seek to minimize the quantity within a finite subspace resulting in quasidiabatic [46, 47] or regularized diabatic states [48, 49]. The adiabatic to diabatic transformation is only defined up to a constant which can be fixed by setting the adiabatic basis equal to the diabatic basis at one specific value of the nuclear coordinates. Thus, the diabatic basis is not unique. In a rigorous diabatic basis, i.e. a basis in which the derivative couplings have been completely removed by the transformation in Eq. 1.15, the time-dependent Schrödinger equation for the nuclear functions is given by

$$i\frac{\partial}{\partial t}\widetilde{\Phi}^{(v)}(\mathbf{R}, t) = \hat{T}_n\widetilde{\Phi}^{(v)}(\mathbf{R}, t) + \sum_w W^{(v,w)}(\mathbf{R})\widetilde{\Phi}^{(w)}(\mathbf{R}, t) \qquad (1.16)$$

Here, the couplings between the electronic states are given by the electronic off-diagonal elements of the diabatic potential coupling $W^{(v,w)}$ which are matrix elements of $\hat{H}_{el}$ in the diabatic basis. The couplings between electronic states in the diabatic representation are thus potential couplings as opposed to the derivative couplings, which depend on the nuclear kinetic energy operator, found in the adiabatic representation. Numerically, the potential couplings are easier to handle why the diabatic representation is often preferred when performing nuclear wavepacket dynamics, however, the adiabatic representation can also be used as we will see in Sect. 4.2. The diabatic potential energy surfaces given by the on-diagonal elements $W^{(v,v)}$ are generally smooth functions of the nuclear coordinates [50, 51] and can preserve the configurational character of the electronic states [52]. In contrast, the adiabatic states preserve the energetic state ordering. To avoid the adiabatic to diabatic transformation, one can also directly construct potential energy surfaces which are smooth functions of the nuclear coordinates as done in Sect. 4.3. This diabatization by ansatz is inherently approximate as the derivative couplings are assumed negligible in the diabatic basis. Nevertheless, the strategy can be useful in particular because the diabatic to adiabatic transformation is unique and is obtained by diagonalization of the matrix representation of the diabatic potential operator to obtain uncoupled states.

1.3 Pump-Probe

We will in this chapter describe the principle underlying pump-probe experiments as a means to interrogate dynamical processes. In particular, we will describe techniques involving short laser pulses in the visible and UV regime where the laser-matter interaction ultimately leads to ionization, and charged particles are detected. The application of these techniques to the investigation of non-adiabatic dynamics will be detailed.

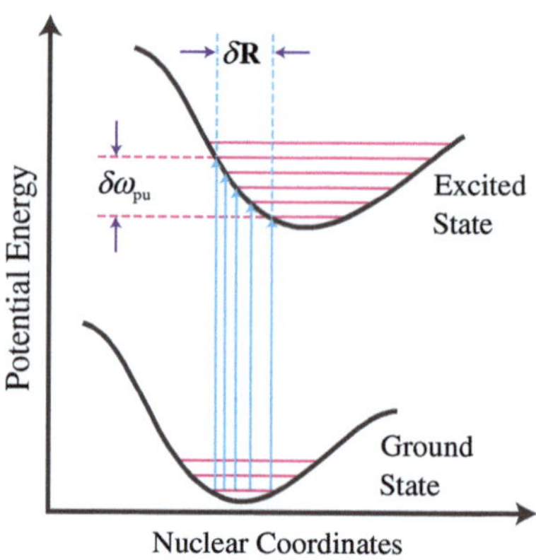

Fig. 1.4 Excitation of the vibrational ground state wavefunction by a short laser pulse leads to a wavepacket in the excited state due to population of a coherent superposition of vibrational eigenstates within the spectral bandwidth of the pump pulse $\delta\omega_{\mathrm{pu}}$. The wavepacket is initially localized to a region $\delta\mathbf{R}$ in nuclear coordinate space

1.3.1 Time-Resolved Photoelectron Spectroscopy and Mass Spectrometry

In a pump-probe experiment, a system initially in a stationary state will through interaction with the pump be taken to a non-stationary state and consequently evolve in time. After a well-defined time-delay, interaction with the probe will interrogate the system to determine the evolution as a function of the time-delay. In time-resolved photoelectron spectroscopy (TRPES) [53–58] and mass spectrometry (TRMS), [4, 59–63] the interaction with the pump pulse leads to electronic excitation of the molecule under investigation. This, in effect, is a projection of the ground state nuclear probability amplitude distribution onto the excited state. Due to the temporal coherence and spectral bandwidth of the laser pulse, this results in the formation of a quantum state localized in real space (Fig. 1.4). By analogy with classical mechanics, this state will start to oscillate on the excited state potential energy surface. From a different point of view, the pump pulse populates a coherent superposition of vibrational eigenstates in the excited state denoted by v

$$|\Phi^{(v)}(t)\rangle = \sum_j C_j^{(v)} e^{-i\omega_j^{(v)}t} |\Phi_j^{(v)}\rangle \qquad (1.17)$$

Due to the differing phase factors of the vibrational eigenstates $|\Phi_j^{(v)}\rangle$, this superposition does not constitute a stationary state but will evolve in time. After a specific time-delay, the interaction with the probe pulse leads to ionization of the molecule. In this process, the wavepacket created by the pump pulse is projected onto the ionization continua of cationic and photoelectron states. TRPES involves measurement of the kinetic energies and angular distributions of the ejected photoelectrons, i.e. it is a frequency-dispersed technique, whereas TRMS involves measurement of the complementary cationic fragments, i.e. it is a frequency-integrated, but mass-dispersed, technique.

1.3.2 Probing Dynamics by TRPES and TRMS

Under certain assumptions, the time-resolved photoelectron spectrum obtained by
ionization of the wavepacket given in Eq. 1.17 to the cationic state w can be given as

$$\sigma(E_k, \Delta t) \propto \int |\mu^{(w,v)}(E_k; \mathbf{R})|^2 \mathcal{F}(\omega_{\mathrm{pr}}, E_k; \mathbf{R})|\Phi^{(v)}(\mathbf{R}, \Delta t)|^2 \, d\mathbf{R} \qquad (1.18)$$

where E_k is the kinetic energy of the ejected photoelectron, Δt is the time between
pump and probe pulses, ω_{pr} is the center frequency of the probe, and $\mathbf{R}$ denote nuclear
coordinates. This expression will be derived in Sect. 5.1. The expression represents
the spectrum as the overlap between three terms: the square of the electronic transition dipole moment, an energy window function, and the density of the nuclear
wavepacket. From this expression, it will be clear how TRPES is sensitive to both
nuclear and electronic dynamics.

When population transfers from the wavepacket in state v to form a wavepacket
in another state v', the density of the nuclear wavepacket in state v decreases. Thus,
population transfer is reflected by a decrease in intensity of bands in the spectrum
as a function of Δt. The simultaneous formation of a new wavepacket in state v'
will be reflected by the appearance of bands, the spectral position and intensity
of which will be determined by the window function and by the transition dipole
moment and nuclear density respectively. The intensity of these new bands will
increase with Δt as more and more population is transferred from state v to state
v'. The situation is exemplified by the schematics in Fig. 1.5 for two distinct cases
which also illustrate the sensitivity of TRPES to the electronic configuration of the
states involved. Employing Koopmans' theorem, [64] different neutral states can
be shown to correlate with different cationic states as depicted in Fig. 1.5(a) – so-
called complementary ionization correlation [65, 66]. These correlations will be
reflected in the magnitude of the electronic transition dipole moment between the
neutral and cationic states with a large magnitude for pairs of states that correlate and
vice versa. Consequently, ionization will primarily occur from a given neutral state
to a specific cationic state and following an electronic transition between neutral
states, the cationic state to which ionization occurs can thus change. The case of
corresponding ionization correlation, where the neutral excited states correlate with
the same cationic state, [16] is shown in Fig. 1.5(b).

To appreciate the sensitivity of TRPES to nuclear dynamics, consider the simple
single trajectory classical limit where the density of the nuclear wavepacket is given
by a δ-function in nuclear coordinate space. In this limit, the spectrum is simply
given by

$$\sigma(E_k, \Delta t) \propto \int |\mu^{(w,v)}(E_k; \mathbf{R})|^2 \mathcal{F}(\omega_{\mathrm{pr}}, E_k; \mathbf{R})|\delta(\mathbf{R} - \mathbf{R}_0(\Delta t)) \, d\mathbf{R}$$

$$= |\mu^{(w,v)}(E_k; \mathbf{R}_0(\Delta t))|^2 \mathcal{F}(\omega_{\mathrm{pr}}, E_k; \mathbf{R}_0(\Delta t)) \qquad (1.19)$$

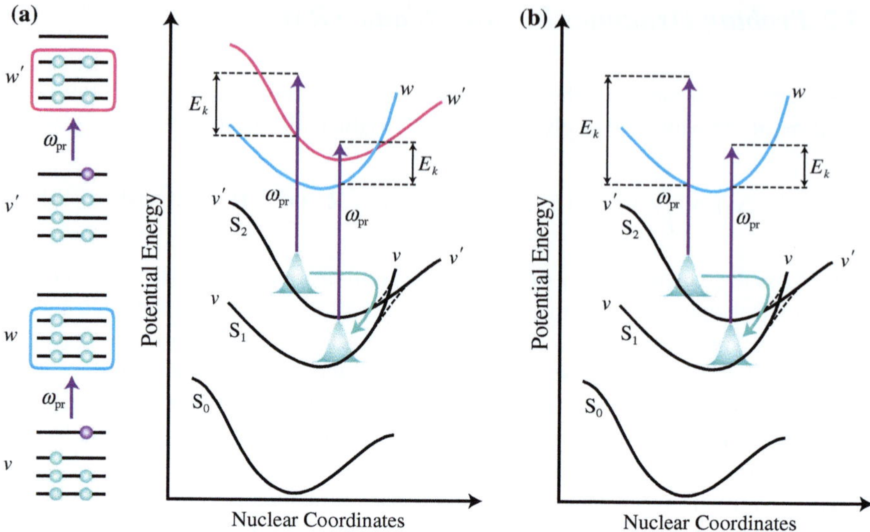

Fig. 1.5 A non-adiabatic transition of a nuclear wavepacket from $S_2 \rightarrow S_1$ is reflected in a change of the kinetic energy of the ejected photoelectron E_k. Conservation of nuclear kinetic energy upon ionization has been assumed in all cases. (**a**) Ionization out of two different neutral excited states v and v' correlating with two different cationic states w and w' respectively. The electronic configurations are shown on the left. (**b**) Ionization out of two different neutral excited states v and v' correlating with the same cationic state w

Thus, nuclear dynamics is reflected in the spectrum through intensity modulations due to the dependence on the nuclear coordinates, which change as a function of Δt, of the electronic transition dipole moment. Furthermore, the dynamics can be reflected in spectral shifts in E_k with Δt due to the energy window function. In one limit, the window function is simply given by

$$\mathcal{F}(\omega_{\mathrm{pr}}, E_k; \mathbf{R}_0(\Delta t)) = \delta(V^{(v)}(\mathbf{R}_0(\Delta t)) - V^{(w)}(\mathbf{R}_0(\Delta t)) - E_k + \omega_{\mathrm{pr}}) \quad (1.20)$$

where $V^{(v)}(\mathbf{R}_0(\Delta t))$ and $V^{(w)}(\mathbf{R}_0(\Delta t))$ are the potential energies at $\mathbf{R}_0(\Delta t)$ of states v and w respectively. The δ-function form clearly reflects the possibility for spectral shifts. The window function given in Eq. 1.20 is valid when nuclear kinetic energy is conserved upon ionization. When kinetic energy is not conserved upon ionization, the photoelectron spectrum can exhibit indistinct features resulting in both nuclear and electronic dynamics being hard to point out.

In favorable cases, TRMS is also sensitive to both nuclear and electronic dynamics. TRMS is a frequency-integrated technique, and the total ion yield is given as the integral of the expression in Eq. 1.18 over photoelectron kinetic energies. Thus, the nuclear and electronic dynamics described above which give rise to changes in the total intensity of the spectrum are also directly reflected in the total ion yield. Furthermore, changes in the fragmentation pattern upon ionization can also directly reveal nuclear [67–69] or electronic dynamics, [70] but these correlations are not necessarily easily deducible. A very direct way of ensuring that an electronic transition is observed by changes in the ion yield is to chose the probe energy such that ionization out of one but not lower-lying electronic states lies within the energy window function. However, ionization by multiple probe photons cannot be distinguished from the one photon case which can obscure results. On the other hand, ionization by different number of probe photons is readily distinguished in TRPES due to the frequency-dispersed detection.

References

1. A.H. Zewail, Angew. Chem. Int. Ed. **39**, 2586–2631 (2000)
2. R.D. Levine, *Molecular Reaction Dynamics* (Cambridge University Press, Cambridge, 2005)
3. N.E. Henriksen, F.Y. Hansen, *Theories of Molecular Reaction Dynamics* (Oxford University Press, Oxford, 2008)
4. N.F. Scherer, J.L. Knee, D.D. Smith, A.H. Zewail, J. Phys. Chem. **89**, 5141–5143 (1985)
5. M. Dantus, M.J. Rosker, A.H. Zewail, J. Chem. Phys. **87**, 2395–2397 (1987)
6. T.S. Rose, M.J. Rosker, A.H. Zewail, J. Chem. Phys. **88**, 6672–6673 (1988)
7. A.H. Zewail, Science **242**, 1645–1653 (1988)
8. R.W. Schoenlein, L.A. Peteanu, R.A. Mathies, C.V. Shank, Science **254**, 412–415 (1991)
9. Q. Wang, R.W. Schoenlein, L.A. Peteanu, R.A. Mathies, C.V. Shank, Science **266**, 422–424 (1994)
10. D. Polli, P. Altoè, O. Weingart, K.M. Spillane, C. Manzoni, D. Brida, G. Tomasello, G. Orlandi, P. Kukura, R.A. Mathies, M. Garavelli, G. Cerullo, Nature **467**, 440–443 (2010)
11. C.E. Crespo-Hernández, B. Cohen, P.M. Hare, B. Kohler, Chem. Rev. **104**, 1977–2019 (2004)
12. C.E. Crespo-Hernández, B. Cohen, B. Kohler, Nature **436**, 1141–1144 (2005)
13. W.J. Schreier, T.E. Schrader, F.O. Koller, P. Gilch, C.E. Crespo-Hernández, V.N. Swaminathan, T. Carell, W. Zinth, B. Kohler, Science **315**, 625–629 (2007)
14. D.R. Cyr, C.C. Hayden, J. Chem. Phys. **104**, 771–774 (1996)
15. E.W.-G. Diau, S. De Feyter, A.H. Zewail, J. Chem. Phys. **110**, 9785–9788 (1999)
16. M. Schmitt, S. Lochbrunner, J.P. Shaffer, J.J. Larsen, M.Z. Zgierski, A. Stolow, J. Chem. Phys. **114**, 1206–1213 (2001)
17. S. Ullrich, T. Schultz, M.Z. Zgierski, A. Stolow, Phys. Chem. Chem. Phys. **6**, 2796–2801 (2004)
18. H. Nakamura, *Nonadiabatic Transition - Concepts*, 2nd edn. (Basic Theories and Applications, World Scientific, Singapore, 2012)
19. R.A. Marcus, N. Sutin, Biochim. Biophys. Acta **811**, 265–322 (1985)
20. A.M. Kuznetsov, J. Ulstrup, *Electron Transfer in Chemistry and Biology - An Introduction to the Theory* (John Wiley & Sons, Chichester, 1999)
21. M. Bixon, J. Jortner, Adv. Chem. Phys. **106**, 35–202 (1999)
22. M.H. Alexander, G. Capecchi, H.-J. Werner, Science **296**, 715–718 (2002)
23. S.C. Althorpe, D.C. Clary, Annu. Rev. Phys. Chem. **54**, 493–529 (2003)
24. L. Che, Z. Ren, X. Wang, W. Dong, D. Dai, X. Wang, D.H. Zhang, X. Yang, L. Sheng, G. Li, H.-J. Werner, F. Lique, M.H. Alexander, Science **317**, 1061–1064 (2007)
25. Y. Huang, C.T. Rettner, D.J. Auerbach, A.M. Wodtke, Science **290**, 111–114 (2000)
26. B. Gergen, H. Nienhaus, W.H. Weinberg, E.W. McFarland, Science **294**, 2521–2523 (2001)
27. J.D. White, J. Chen, D. Matsiev, D.J. Auerbach, A.M. Wodtke, Nature **433**, 503–505 (2005)
28. G.A. Somorjai, *Introduction to Surface Chemistry and Catalysis* (John Wiley & Sons, New York, NY, 1994)
29. G. Ertl, Angew. Chem. Int. Ed. **47**, 3524–3535 (2008)

30. T. Baer, W.L. Hase, *Unimolecular Reaction Dynamics - Theory and Experments* (Oxford University Press, New York, NY, 1996)
31. D.A. McQuarrie, *Statistical Mechanics* (University Science Books, Sausalito, CA, 2000)
32. E.W.-G. Diau, J.L. Herek, Z.H. Kim, A.H. Zewail, Science **279**, 847–851 (1998)
33. K. B. Møller and A. H. Zewail, Femtosecond Activation of Reactions: The Concepts of Non-ergodic Behavior and Reduced-Space Dynamics, in Essays in Contemporary Chemistry: From Molecular Structure towards Biology, edited by G. Quinkert and M. V. Kisakürek, chapter 5, pages 157–188, Verlag Helvetica Chimica Acta, Zürich, and Wiley-VCH, Weinheim, 2001.
34. C.J. Ballhausen, A.E. Hansen, Annu. Rev. Phys. Chem. **23**, 15–38 (1972)
35. W. Domcke, G. Stock, Adv. Chem. Phys. **100**, 1–169 (1997)
36. G.A. Worth, L.S. Cederbaum, Annu. Rev. Phys. Chem. **55**, 127–158 (2004)
37. M. Born, K. Huang, *Dynamical Theory of Crystal Lattices* (Oxford University Press, London, 1954)
38. M. Born, R. Oppenheimer, Ann. Phys. (Leipzig) **84**, 457–484 (1927)
39. V. Sidis, Adv. Chem. Phys. **82**, 73–134 (1992)
40. D.R. Yarkony, Rev. Mod. Phys. **68**, 985–1013 (1996)
41. D.R. Yarkony, Acc. Chem. Res. **31**, 511–518 (1998)
42. D.R. Yarkony, J. Phys. Chem. A **105**, 6277–6293 (2001)
43. T.J. Martínez, Nature **467**, 412–413 (2010)
44. G.J. Atchity, S.S. Xantheas, K. Ruedenberg, J. Chem. Phys. **95**, 1862–1876 (1991)
45. C.A. Mead, D.G. Truhlar, J. Chem. Phys. **77**, 6090–6098 (1982)
46. T. Pacher, L.S. Cederbaum, H. Köppel, Adv. Chem. Phys. **84**, 293–391 (1993)
47. M. Baer, Phys. Rep. **358**, 75–142 (2002)
48. H. Köppel, J. Gronki, S. Mahapatra, J. Chem. Phys. **115**, 2377–2388 (2001)
49. H. Köppel, B. Schubert, Mol. Phys. **104**, 1069–1079 (2006)
50. F.T. Smith, Phys. Rev. **179**, 111–123 (1969)
51. R.D. Levine, B.R. Johnson, R.B. Bernstein, J. Chem. Phys. **50**, 1694–1701 (1969)
52. H. Nakamura, D.G. Truhlar, J. Chem. Phys. **115**, 10353–10372 (2001)
53. R. Haight, J. Bokor, J. Stark, R.H. Storz, R.R. Freeman, P.H. Bucksbaum, Phys. Rev. Lett. **54**, 1302–1305 (1985)
54. D.M. Neumark, Annu. Rev. Phys. Chem. **52**, 255–277 (2001)
55. A. Stolow, Annu. Rev. Phys. Chem. **54**, 89–119 (2003)
56. A. Stolow, A.E. Bragg, D.M. Neumark, Chem. Rev. **104**, 1719–1757 (2004)
57. T. Suzuki, Annu. Rev. Phys. Chem. **57**, 555–592 (2006)
58. A. Stolow, J.G. Underwood, Adv. Chem. Phys. **139**, 497–583 (2008)
59. J.W. Perry, N.F. Scherer, A.H. Zewail, Chem. Phys. Lett. **103**, 1–8 (1983)
60. N.F. Scherer, J.F. Shepanski, A.H. Zewail, J. Chem. Phys. **81**, 2181–2182 (1984)
61. J.L. Knee, F.E. Doany, A.H. Zewail, J. Chem. Phys. **82**, 1042–1043 (1985)
62. J.L. Knee, L.R. Khundkar, A.H. Zewail, J. Chem. Phys. **82**, 4715–4716 (1985)
63. N.F. Scherer, L.R. Khundkar, T.S. Rose, A.H. Zewail, J. Phys. Chem. **91**, 6478–6483 (1987)
64. T. Koopmans, Physica **1**, 104–113 (1933)
65. V. Blanchet, M.Z. Zgierski, T. Seideman, A. Stolow, Nature **401**, 52–54 (1999)
66. V. Blanchet, M.Z. Zgierski, A. Stolow, J. Chem. Phys. **114**, 1194–1205 (2001)
67. C. Kötting, E.W.-G. Diau, T.I. Sølling, A.H. Zewail, J. Phys. Chem. A **106**, 7530–7546 (2002)
68. J.-W. Ho, W.-K. Chen, P.-Y. Cheng, J. Chem. Phys. **131**, 134308 (2009)
69. R.Y. Brogaard, K.B. Møller, T.I. Sølling, J. Phys. Chem. A **115**, 12120–12125 (2011)
70. J.L. Gosselin, M.P. Minitti, F.M. Rudakov, T.I. Sølling, P.M. Weber, J. Phys. Chem. A **110**, 4251–4255 (2006)

Part II
Experimental Methods

Chapter 2
Experimental Setup

We will in this chapter describe the experimental setup for conducting time-resolved photoionization experiments such as time-resolved mass spectrometry (TRMS) and photoelectron spectroscopy (TRPES) on gaseous samples. The main setup consists of a laser system capable of producing laser pulses of a femtosecond time duration, a means of preparing a cold gaseous sample, and a detection apparatus for charged particles. We will in the following sections describe these components.

2.1 The Laser System and Optical Setup

The laser system and optical setup used in the experiments is schematically illustrated in Fig. 2.1. The laser system consists of a Ti:Sapphire oscillator (Spectra Physics Tsunami, 80 MHz, 800 nm, 80 fs, 700 mW) which is pumped by a Nd:YLF continuous wave (CW) laser (Spectra Physics Millennia Pro Vs, 532 nm, 5 W). The oscillator seeds a regenerative Ti:Sapphire amplifier (Spectra Physics SpitFire, 1 kHz, 800 nm, 100 fs, 1.1 W) which is pumped by a Q-switched $Nd:YVO_4$ laser (Spectra Physics Empower-15, 1 kHz, 527 nm, 100 ns, 6.7 W). The output of the amplifier is split and used to pump an optical parametric amplifier (Light Conversion TOPAS-C, 240–2600 nm) and a harmonic generation setup which can generate the second (2ω), third (3ω), and fourth harmonic (4ω) of the fundamental. In addition to the laser system, the setup also consists of a retro-reflector mounted on a computer-controlled translatable stage on the harmonic generation arm of the optical setup. The stage allows for changes to the path length of the harmonic generation arm and can thereby be used to change the time-delay between pulses from the two arms. The beams from the two arms are combined using a dichroic mirror, and the collinear beams are focused into the vacuum chamber using a concave aluminium mirror. Pulses from either arm can act as the pump pulses in the experiments, and pulses from the other arm will act as the probes.

T. S. Kuhlman, *The Non-Ergodic Nature of Internal Conversion*, Springer Theses,
DOI: 10.1007/978-3-319-00386-3_2, © Springer International Publishing Switzerland 2013

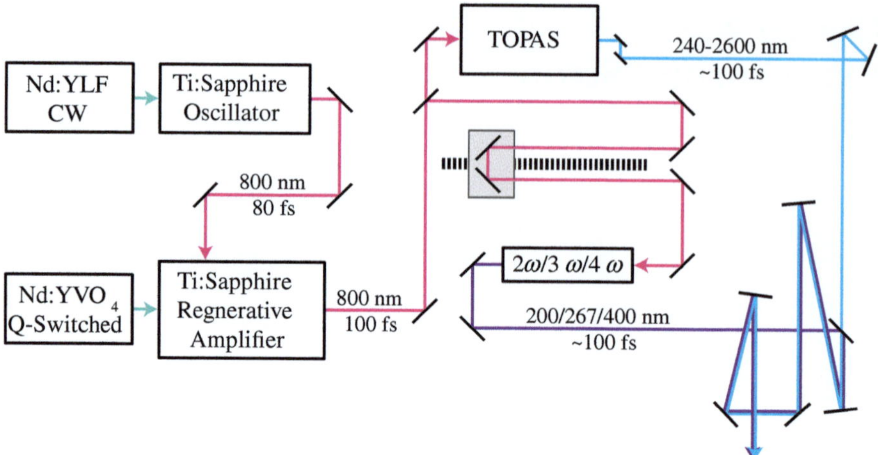

Fig. 2.1 Schematic overview of the laser system and optical setup including the stage used for controlling the time-delay between the two laser pulses. Wavelengths and pulse durations are indicated

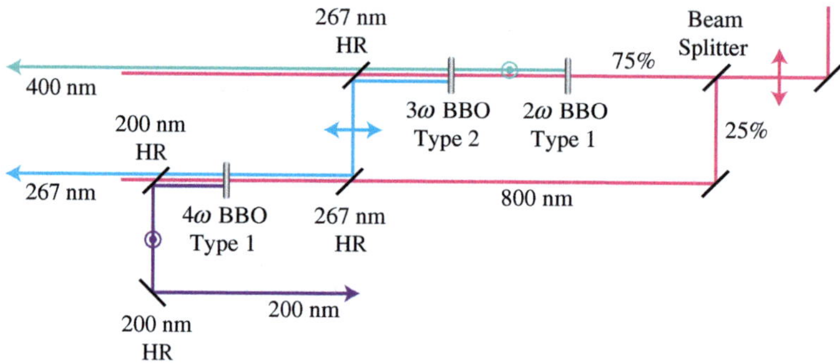

Fig. 2.2 Schematic overview of the 2ω, 3ω, and 4ω generation setup. Polarization vectors are indicated for the different harmonics, and the important optical elements are labeled

The harmonic generation setup is illustrated schematically in Fig. 2.2. The fundamental (800 nm) is split 25/75 in a beam splitter, and the larger fraction is frequency doubled by type 1 second-harmonic generation (SHG) in a thin beta barium borate (BBO) crystal. In a second BBO crystal, the co-propagating fundamental and second harmonic (400 nm) create the third harmonic (267 nm) by type 2 sum-frequency generation (SFG). Two dichroic mirrors (267 nm HR) reflect the third harmonic towards the third BBO crystal in which it combines with the smaller fraction of the fundamental split off by the beam splitter to create the fourth harmonic (200 nm) by type 1 SFG. This third BBO crystal is uncoated to avoid damage due to the UV radiation. As BBO is hygroscopic, the crystal is constantly heated to 50 °C. The fourth harmonic is reflected by two dichroic mirrors (200 nm HR) to split off the remaining fundamental and third harmonic. The third harmonic can easily be extracted from the setup by removing the beam splitter and the first 267 nm HR whereas the second harmonic can be extracted if the second BBO crystal is also removed.

2.2 Time-of-Flight Spectrometers

The vacuum chamber incorporating the time-of-flight (TOF) spectrometers as well as the molecular beam generation is illustrated in Fig. 2.3. The setup can handle both gaseous, liquid, and solid samples. In the case of liquid samples, the inert helium carrier gas is bubbled through the sample at a backing pressure of 0.3 bar, and a mixture of helium and sample molecules are let into the chamber through a stainless steel inlet tube. Solid samples can be placed in a small heatable chamber at the end of the inlet tube and sublimed into the helium carrier gas.

In the first chamber, the gas mixture continuously expands through a conical nozzle ($\varnothing = 100\,\mu$m) at the end of the inlet tube. In the course of the expansion, the sample molecules are first accelerated to the velocity of the carrier gas whereafter they start transferring energy to the carrier gas molecules through two-body interactions thereby cooling their rotational and vibrational degrees of freedom [1–3]. Rotational temperatures down to $\sim$2 K can be obtained with continuous inlet systems such as the one in our setup whereas high-pressure pulsed valve systems can obtain temperatures down to $\sim$0.4 K [3, 4]. The expansion beam is termed a supersonic molecular beam as the speed of the carrier molecules ($\sim$1750 m/s for He at room temperature [3]) can exceed that of sound. The expansion of the mixture leads to a low-density sample thereby eliminating intermolecular interactions providing an ensemble of molecules in one or a few well-defined quantum states [5]. Approximately one centimeter from the nozzle of the inlet tube, a skimmer ($\varnothing = 200\,\mu$m) ensures the formation of a turbulence free molecular beam in the next chamber. From the middle chamber, the beam passes through a pinhole ($\varnothing = 1$ mm) into the main interaction chamber where it is intersected at a right angle by the laser beams.

The interaction between the laser pulses and the beam molecules leads to photoionization forming cations and photoelectrons. These charged particles can be extracted and detected at the top or bottom of the chamber by the respective detector. Thus, the apparatus operates as either a TOF mass or photoelectron spectrometer but cannot be operated in coincidence. The mass spectrometer can also be operated as a reflectron TOF mass spectrometer although this mode of operation was not employed in this work.

When operated as a mass spectrometer, the apparatus uses a two-field Wiley-McLaren configuration [6]. The voltage difference $V_E = V_{A1} - V_{A2}$ creates a uniform field $\mathcal{E}_E = -V_E/l_E$ as shown in the inset of Fig. 2.3. This field extracts the cations generated by the laser-molecule interaction. The voltage difference $V_A = V_{A2} - V_L$ creates another uniform field $\mathcal{E}_A = -V_A/l_A$ which accelerates the cations into the field free flight tube above the accelerator grid. At the top of the flight tube, the cations are incident upon the ion detector consisting of a set of chevron-stacked

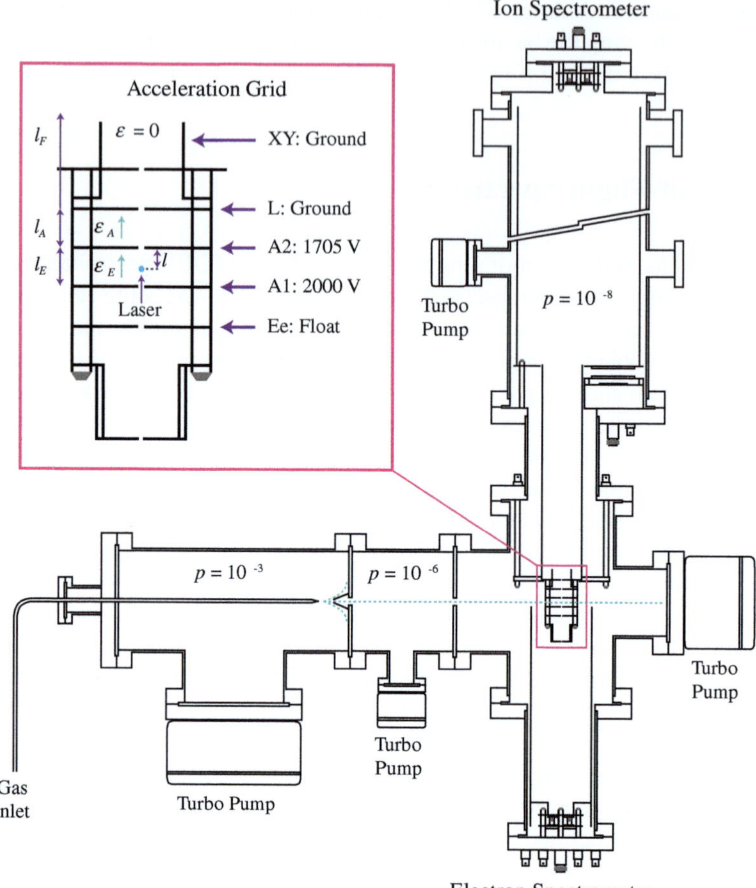

Fig. 2.3 Schematic overview of the vacuum chamber, molecular beam generation, and TOF spectrometers. The inset shows a close-up of the acceleration grid with typical voltages employed when the apparatus is operated in ion mode. Measurements run in electron mode are field free, and all grid connections are grounded. The typical pressures in the chambers (mbar) are indicated showing the differential pumping. Dimensions are not to scale

micro-channel plates (MCP) operated at negative potential. A small DC voltage ($\sim 70\,\mathrm{mV}$) is generated in an anode on the backside of the MCP stack when an ion impinges on the detector. This voltage signal is subsequently sent to a time digitizer card (FAST ComTec P7888-2(E)) in a computer. As the card also receives a trigger signal from the laser system, the computer can register the TOF for the impinging cations. The total TOF of the cations is given by the sum of the times spent in the extraction, acceleration, and free flight regions which are given by [7]

$$t_E = \sqrt{\frac{m}{z}}\sqrt{\frac{2l_E l}{V_E}} \tag{2.1}$$

$$t_A = l_A \sqrt{\frac{m}{z}} \left[\sqrt{\frac{2}{V_A}\left(1 + \frac{V_E l}{V_A l_E}\right)} - \sqrt{\frac{2}{V_A}\frac{V_E l}{V_A l_E}} \right] \tag{2.2}$$

$$t_F = l_F \sqrt{\frac{m}{z}}\sqrt{\frac{1}{2V_A + 2V_E l/l_E}} \tag{2.3}$$

It is assumed that all photoionization events occur at the same distance l from the acceleration region, and that the initial velocity along the flight tube axis is zero. The total TOF is dependent on the mass to charge ratio m/z, and the cations are therefore separated into bunches when they reach the detector with the lightest first followed by successively heavier masses (assuming identical charges). Taking into account that the digitizer card records the TOF with respect to the laser trigger signal, which is not identical to the time of photoionization, the masses of the recorded signal are given by

$$m/z = K_{\mathrm{ion}}(t_{\mathrm{ion}}^{\mathrm{TOF}} - t_0)^2 \tag{2.4}$$

where all the spectrometer dependent parameters are collected into a single constant K_{ion}, and t_0 measures the difference between the laser trigger and the time of photoionization. The mass spectrometer is calibrated by measuring samples with know mass spectra usually acetone, N, N-dimethylisopropylamine (DMIPA), and xenon.

When operated as a photoelectron spectrometer, all connections on the acceleration grid are grounded. A tube of μ-metal shields the inside of the downward flight tube from magnetic fields, and the experiments are (ideally) run field free. Furthermore, the surfaces of the accelerator are made from oxidized molybdenum which has a high work function such as to reduce background photoelectrons. Nonetheless, scattered photons can contribute to the background signal, and, therefore, an iris is placed in the vacuum chamber before the interaction region, and a laser-grade CaF_2 entrance window is used (CVI PW-1009-CFUV). The part of the photoelectrons generated by the laser-molecule interaction with a primarily downward pointing velocity vector will be incident upon the electron detector. This detector consists of another set of chevron-stacked MCPs operated at a positive potential. The anode is also operated at a high positive potential, and the signal from the electrons is thus at a high DC potential. The timing electronics are protected by a capacitive decoupling of the anode. The decoupled, negative DC voltage signal ($\sim$10 mV) is passed through a pre-amplifier (Phillips Scientific 6954B 100), and the amplified signal ($\sim$500 mV) is sent to the time digitizer card. Similar to the case of ions, the TOF is measured which in this case is given by

$$t_e^{\mathrm{TOF}} = l_F \sqrt{\frac{1}{2(E_k + E_0)}} \tag{2.5}$$

where E_k is the kinetic energy gained by the photoelectrons in the ionization process. E_0 is a correction term resulting from deviations from field free operation due to e.g. contact potential terms leading to acceleration/decelleration of the photoelectrons. The latter term is usually on the order of meV. l_F is here the length of the downward free flight path of the photoelectrons. Inverting the expression in Eq. 2.5 yields the kinetic energy as a function of the TOF

$$E_k = \frac{K_e}{\left(t_e^{\text{TOF}} - t_0\right)^2} + E_0 \tag{2.6}$$

We have again collected parameters and constants into a single coefficient K_e, and the time difference between trigger and photoionization is taken into account via t_0. Equation 2.6 provides the position of the spectral peaks, however, to determine the full photoelectron spectrum, the Jacobian has to be taken into account to establish the correct relative spectral intensities

$$\sigma(E_k) = \sigma(E_k(t_e^{\text{TOF}})) \left| \frac{\mathrm{d}t_e^{\text{TOF}}}{\mathrm{d}E_k} \right| = \sigma(E_k(t_e^{\text{TOF}})) \frac{\left(t_e^{\text{TOF}}\right)^3}{2K_e} \tag{2.7}$$

When determining the mass spectrum, it is not necessary to take the Jacobian into account as the cationic signals are only spread over a few TOF bins, i.e. the spectrum consists of almost discrete peaks. Similarly to the case for ions, the parameters K_e, E_0 and t_0 are determined from TOF spectra with sharp peaks of known energy usually xenon and DMIPA.

The above procedures describe how to obtain mass and photoelectron spectra $\sigma(m/z)$ and $\sigma(E_k)$ from the appropriate TOF spectra $\sigma(t_{\text{ion}}^{\text{TOF}})$ and $\sigma(t_e^{\text{TOF}})$. A Lab-View program records the data from the time digitizer card and furthermore controls the retroreflector stage. By repeating the above procedures for a given set of pump-probe time-delays by moving the stage, time-resolved spectra $\sigma(t_{\text{ion}}^{\text{TOF}}, \Delta t)$ and $\sigma(t_e^{\text{TOF}}, \Delta t)$ can automatically be recorded by the program from which the spectra $\sigma(m/z, \Delta t)$ and $\sigma(E_k, \Delta t)$ can be determined.

References

1. G. Scoles (ed.), *Atomic and Molecular Beam Methods*, vol. 1 (Oxford University Press, New York, 1988)
2. G. Scoles (ed.), *Atomic and Molecular Beam Methods*, vol. 2 (Oxford University Press, New York, 1992)
3. U. Even, J. Jortner, D. Noy, N. Lavie, C. Cossart-Magos, J. Chem. Phys. **112**, 8068–8071 (2000)
4. A. Amirav, U. Even, J. Jortner, Chem. Phys. **51**, 31–42 (1980)
5. V. Vaida, Acc. Chem. Res. **19**, 114–120 (1986)
6. W.C. Wiley, I.H. McLaren, Rev. Sci. Instrum. **26**, 1150–1157 (1955)
7. F. Chandezon, B. Huber, C. Ristori, Rev. Sci. Instrum. **65**, 3344–3353 (1994)

Chapter 3
Fitting of Experimental Data and Cross-Correlation

The time-dependent spectra obtained by the procedures described in the previous chapter provide a wealth of information on the dynamical processes of the molecules investigated. To derive manageable data from the spectra, it is often assumed that the individual dynamical steps investigated follow first-order kinetics [1]. In the experiments, the time-resolution is limited by the instrument response function which is dominated by the temporal cross-correlation (XC) between the pump and probe pulses. Therefore, the spectra are fitted to functions of the form

$$\sigma(\Delta t) = \vartheta_{\mathrm{xc}}(\Delta t) \otimes \Theta(\Delta t) \sum_j A_j \vartheta_j(\Delta t) \tag{3.1}$$

Here, $\vartheta_{\mathrm{xc}}(\Delta t)$ is the XC component, $\Theta(\Delta t)$ is the Heaviside step function, $\vartheta_j(\Delta t)$ is the solution to a given kinetic model with amplitude A_j, and $\otimes$ represents a convolution. When Eq. (3.1) is used for fitting full two-dimensional spectra, the amplitudes A_j are spectral components whereas when fitting one-dimensional transients, they are simply scalars. The XC is assumed to be a Gaussian function. As an example, a single first-order process is described by the form

$$\sigma(\Delta t) = \exp\left[-4\ln(2)\Delta t^2/\tau_{\mathrm{xc}}^2\right] \otimes A_1 \exp\left[-\Delta t/\tau_1\right]\Theta(\Delta t) \tag{3.2}$$

where τ_{xc} is the full width at half maximum (FWHM) of the XC, and τ_1 is the time constant for the exponential decay also referred to as a lifetime. If $\tau_1 \gg \tau_{\mathrm{xc}}$, the signal for times $\Delta t \ll \tau_1$ is simply given by a constant amplitude. In this case, the XC component can be determined from

$$\sigma(\Delta t) = \exp\left[-\left(\frac{2\sqrt{\ln(2)}}{\tau_{\mathrm{xc}}}\right)^2 \Delta t^2\right] \otimes A_1\Theta(\Delta t)$$

$$= A_1 \frac{\tau_{\mathrm{xc}}}{2\sqrt{\ln(2)}}\sqrt{\frac{\pi}{4}}\mathrm{erfc}\left[-\frac{2\sqrt{\ln(2)}}{\tau_{\mathrm{xc}}}\Delta t\right] \tag{3.3}$$

T. S. Kuhlman, *The Non-Ergodic Nature of Internal Conversion*, Springer Theses,
DOI: 10.1007/978-3-319-00386-3_3, © Springer International Publishing Switzerland 2013

Scheme 3.1 Example of sequential first-order processes where the two first species give rise to the same experimental spectrum but with different amplitudes

$$\mathrm{M}_1^* \xrightarrow{\tau_1} \mathrm{M}_2^* \xrightarrow{\tau_2} \mathrm{M}_3^*$$

$$\downarrow \qquad\qquad \downarrow \qquad\qquad \cancel{\downarrow}$$

$$A_1\sigma(\mathrm{M}^*) \qquad\qquad A_2\sigma(\mathrm{M}^*)$$

where erfc is the complementary error function. This approach is usually employed when determining the XC by use of N,N-dimethylisopropylamine. If on the other hand the system investigated does not follow first-order kinetics but exhibits an impulsive response, i.e. a δ-function response, the signal will fully be given by the XC component multiplied by the amplitude of the impulsive response. Such a response can be observed if the multi-photon ionization by the pump and probe pulses is non-resonant, and such a measurement can thereby also be used to determine the XC. This approach is employed using xenon. Non-resonant components can generally occur in experiments even if the pump is chosen to be resonant with a transition in the sample molecules as contributions from non-resonant probe-pump ionization, i.e. probe preceding pump, can be present. This contribution can be added to Eq. (3.1) to obtain

$$\sigma(\Delta t) = A_{\mathrm{xc}}\vartheta_{\mathrm{xc}}(\Delta t) + \vartheta_{\mathrm{xc}}(\Delta t) \otimes \Theta(\Delta t) \sum_j A_j\vartheta_j(\Delta t) \tag{3.4}$$

More complex solutions than a single exponential decay can also arise from first-order kinetics. As an example, this can occur if the process investigated follows sequential first-order kinetics such as $\mathrm{M}_1^* \to \mathrm{M}_2^* \to \mathrm{M}_3^*$, and the two species M_1^* and M_2^* give rise to the same signal $\sigma(\mathrm{M}^*)$ but with different amplitudes. This situation is shown in Scheme 3.1. In this case, the signal is given by a sequential biexponential decay

$$\sigma(\Delta t) = \exp\left[-4\ln(2)\Delta t^2/\tau_{\mathrm{xc}}^2\right]$$
$$\otimes (A_1 \exp[-\Delta t/\tau_1] + A_2 (1 - \exp[-\Delta t/\tau_1]) \exp[-\Delta t/\tau_2]) \,\Theta(\Delta t) \tag{3.5}$$

As another example, if the decay of a given species follows first-order kinetics, but the probability of observing the species is modulated by a periodic perturbation, a suitable functional form is

$$\sigma(\Delta t) = \exp\left[-4\ln(2)\Delta t^2/\tau_{\mathrm{xc}}^2\right]$$
$$\otimes \exp[-\Delta t/\tau_1] (A_1 + A_2 \cos[2\pi\Delta t/T + \phi]) \,\Theta(\Delta t) \tag{3.6}$$

where A_2 is the amplitude of the perturbation with period T and phase ϕ. We will later show how electronic population decay modulated by nuclear motion can be fitted using such a model. The different functional forms are exhibited in Fig. 3.1.

Experimental data are fitted to the models described above by non-linear least-squares analysis, i.e. by minimizing the χ^2 measure [2]

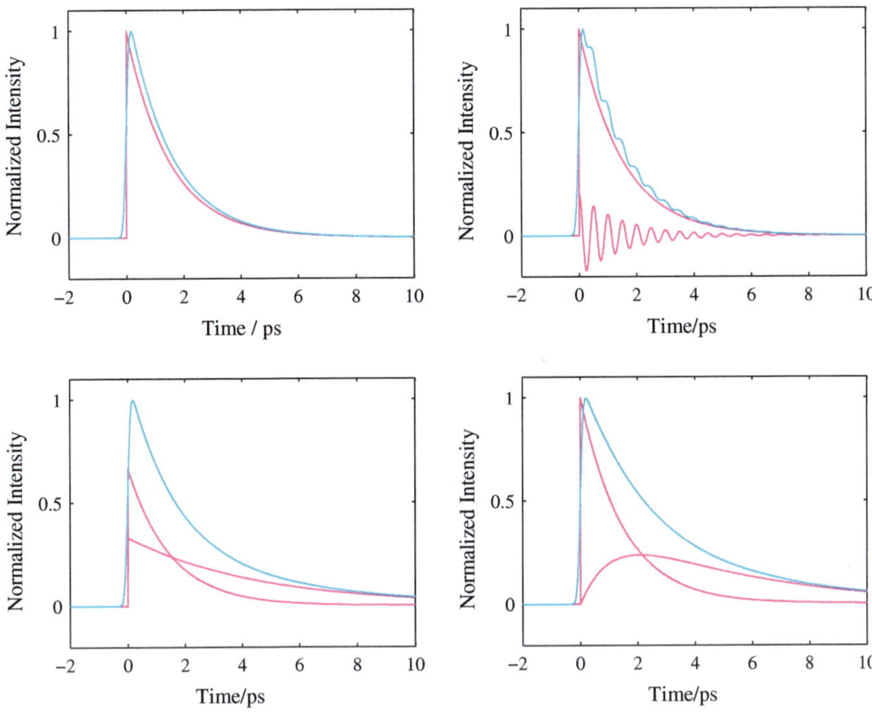

Fig. 3.1 Examples of models used for fitting experimental data (blue) showing the individual components (pink). The time constants τ_1 and τ_2 have been set to 1.5 and 4.5 ps respectively, and $A_1/A_2 = 2$ except for the modulated exponential decay where $A_1/A_2 = 5$. The period of the modulation $T = 0.5$ ps. The models have been convoluted by a XC with $\tau_{xc} = 0.2$ ps. (**a**) Exponential decay. (**b**) Modulated exponential decay. (**c**) Biexponential decay. (**d**) Sequential biexponential decay

$$\chi^2 = \sum_i \left(\frac{r_i}{s_i}\right)^2 \approx \sum_i \frac{r_i^2}{c_i} \tag{3.7}$$

Here, r_i is the residual in the ith channel, and s_i is the standard deviation of the value in channel i. Time-of-flight mass spectrometry and photoelectron spectroscopy result from counting discrete events and follow Poisson statistics. If the number of counts is large enough, the Poisson distribution can be approximated by a normal distribution, and the standard deviation is simply given by the square root of the number of counts in the channel, i.e. $s_i = \sqrt{c_i}$. The global optimization of χ^2 is performed using either a Levenberg-Marquardt [3, 4] or simulated annealing algorithm [5, 6] as implemented in Matlab. Generally, simulated annealing outperforms Levenberg-Marquardt in particular in cases with many parameters or noisy data.

The χ^2 measure can also be used to determine statistical uncertainties in the fitted parameters. Having obtained the parameters that minimize χ^2 to the value of $\chi^2_{\min}$, the standard deviation in a given parameter p can be calculated using [2]

$$s_p = \frac{\Delta p}{\sqrt{\chi^2 - \chi^2_{\min}}} \tag{3.8}$$

Thus, the standard deviation is calculated by changing the value of the given parameter by 1–2 % from the optimal value, i.e. by Δp, and performing a new minimization of χ^2 with the parameter fixed. Inserting the new value of χ^2 into the above equation yields the standard deviation. It should be noted that this procedure only provides statistical uncertainties whereas systematic uncertainty, e.g. due to the specific functional form of the fit, is not reflected in the value.

References

1. A.A. Istratov, O.F. Vyvenko, Rev. Sci. Instrum. **70**, 1233–1257 (1999)
2. P.J. Cumpson, M.P. Seah, Surf. Interface Anal. **18**, 345–360 (1992)
3. K. Levenberg, Q. Appl. Math. **2**, 164–168 (1944)
4. D.W. Marquardt, J. Soc. Ind. Appl. Math. **11**, 431–441 (1963)
5. S. Kirkpatrick, C.D. Gelatt, M.P. Vecchi, Science **220**, 671–680 (1983)
6. V. Černý, J. Optim. Theory Appl. **45**, 41–51 (1985)

Part III
Theoretical Methods

Chapter 4
Nuclear Dynamics

Many methods exist for simulating nuclear dynamics ranging from classical trajectory to full quantum wavepacket methods. When dealing with dynamics on several electronic states, some method of describing transfer between states has to be included. This requirement excludes the use of purely classical methods, and one has to resort to (approximately) solving the time-dependent Schrödinger equation

$$i\frac{\partial}{\partial t}\Psi(\mathbf{r}, \mathbf{R}, t) = \hat{H}\Psi(\mathbf{r}, \mathbf{R}, t) \tag{4.1}$$

with $\mathbf{r}$ and $\mathbf{R}$ representing electronic and nuclear coordinates respectively. In order to accomplish such a task, a suitable representation of the wavefunction $\Psi(\mathbf{r}, \mathbf{R}, t)$ as well as a method to propagate it in time is needed. The total wavefunction is usually given as a product of an electronic and a vibrational wavefunction as described in Sect. 1.2. With a complete set of diabatic or adiabatic electronic states, what remains is to specify a representation of the nuclear wavefunction $\Phi(\mathbf{R}, t)$. Two possible approaches are

$$\Phi(\mathbf{R}, t) = \sum_j C_j(t)\phi_j(\mathbf{R}) \tag{4.2}$$

$$\Phi(\mathbf{R}, t) = \sum_j C_j(t)\phi_j(\mathbf{R}, t) \tag{4.3}$$

In the first approach, the time-dependence is fully contained in the coefficients, and the wavefunction is expanded in a set of time-independent basis functions. These functions are usually chosen as an appropriate set such as harmonic oscillator or Morse eigenfunctions for vibrational degrees of freedom. Pseudo-spectral methods involving a discrete variable representation (DVR) [1] are also of this form [1, 2]. In methods employing a DVR, a real-space grid, on which the wavefunction can be represented, is defined by means of an orthonormal set of basis functions. In

T. S. Kuhlman, *The Non-Ergodic Nature of Internal Conversion*, Springer Theses, 29
DOI: 10.1007/978-3-319-00386-3_4, © Springer International Publishing Switzerland 2013

this way, the first approach also encompasses grid-based methods such as those of Feit and Fleck [3, 4] and Kosloff [5, 6]. We will generally refer to methods employing time-independent basis functions as numerically exact grid-based methods.

In the other approach, increased flexibility in the description of the wavefunction is obtained by allowing both coefficients and basis functions to be time-dependent. In this way, the numerical intractability of grid-based methods for high-dimensional systems can be somewhat alleviated. Several methods in this group can be viewed as extensions of Heller's semi-classical Gaussian wavepackets [7–10]. Such methods include the quasi-classical Full Multiple Spawning (FMS) [11–14], Ab Initio Multiple Spawning (AIMS) [15, 16], and Coupled Coherent States [17–20] as well as the full-quantum method of variational Multi-Configuration Gaussian wavepacket (vMCG) [21, 22]. vMCG is a derivative of Multi-Configuration Time-Dependent Hartree (MCTDH) [23–25]. Unlike vMCG, MCTDH uses general, not parameterized, time-dependent basis functions.

In this chapter, we will give a short introduction to the formalism of MCTDH and that of the Vibronic Coupling Hamiltonian (VCHAM) which were used in conjunction for the simulations presented in Chap. 7. We will show how complex Gaussian functions can be used in the MCTDH formalism leading to the vMCG method. This provides a natural transition to a presentation of the formalism and applicational aspects of the related Gaussian wavepacket methods of FMS and AIMS. AIMS was used for the simulations presented in Chap. 8.

4.1 Multi-Configuration Time-Dependent Hartree

4.1.1 The MCTDH Equations of Motion

In multi-state MCTDH, the total nuclear wavefunction is written as a sum over N_e electronic states according to [26]

$$\Phi(\mathbf{Q}, t) = \sum_{v}^{N_e} \Phi^{(v)}(\mathbf{Q}, t) \tag{4.4}$$

Here, $\Phi^{(v)}(\mathbf{Q}, t)$ is the nuclear wavefunction of electronic state v, and $\mathbf{Q}$ denotes general nuclear coordinates. The diabatic representation is usually employed for the electronic states to avoid having to deal with the possibly diverging derivative coupling terms present in the adiabatic representation (see Sect. 1.2). The ansatz for the nuclear wavefunction takes on the following form [23, 24, 27]

$$\Phi^{(v)}(\mathbf{Q}, t) = \sum_j^{n^{(v)}} C_j^{(v)}(t) \phi_j^{(v)}(\mathbf{Q}, t)$$

$$= \sum_{j_1}^{n_1^{(v)}} \cdots \sum_{j_N}^{n_N^{(v)}} C_{j_1 \cdots j_N}^{(v)}(t) \prod_\kappa^N \varphi_{j_\kappa}^{(v)}(Q_\kappa, t) \tag{4.5}$$

Thus, the wavefunction is given as a sum of Hartree products or configurations $\phi_j^{(v)}(\mathbf{Q}, t)$ giving rise to the name of the method. The Hartree products each carry a complex time-dependent coefficient $C_{j_1 \cdots j_N}^{(v)}(t)$ in the wavefunction expansion, and they are themselves products of N single-particle functions $\varphi_{j_\kappa}^{(v)}(Q_\kappa, t)$. To ease notation, the composite index $j = j_1 \cdots j_N$ has been introduced. The equations of motion for the time-dependent coefficients and single-particle functions are obtained by use of the Dirac-Frenkel variational principle [28, p. 253]

$$\langle \delta\Phi | \hat{H} - i\, (\partial/\partial t) | \Phi \rangle = 0 \tag{4.6}$$

The reader is referred to Ref. [24] for a full derivation of the equations of motion. For the coefficients, the equation reads

$$\dot{\mathbf{C}} = -i\mathbf{H}\mathbf{C} \tag{4.7}$$

where we have introduced the vector of coefficients $\mathbf{C}$ and its time-derivative $\dot{\mathbf{C}}$ as well as the Hamiltonian matrix $\mathbf{H}$ represented in the basis of Hartree products. Explicitly for the coefficients of a given electronic state, the equation reads

$$\dot{\mathbf{C}}^{(v)} = -i \sum_w^{N_e} \mathbf{H}^{(v,w)} \mathbf{C}^{(w)} \tag{4.8}$$

The working equation for the single-particle functions is slightly more involved

$$i\dot{\boldsymbol{\varphi}}_\kappa^{(v)} = (1 - P_\kappa^{(v)}) \left(\rho_\kappa^{(v)} \right)^{-1} \sum_w^{N_e} \widetilde{\mathbf{H}}_\kappa^{(v,w)} \boldsymbol{\varphi}_\kappa^{(w)} \tag{4.9}$$

Here, the vector of single-particle functions $\boldsymbol{\varphi}_\kappa^{(v)}$ for degree of freedom κ and electronic state v and its time-derivative $\dot{\boldsymbol{\varphi}}_\kappa^{(v)}$ have been introduced. We have also introduced three other objects: the projection operator onto the space of single-particle functions for the κth degree of freedom $P_\kappa^{(v)}$ as well as the density matrix $\rho_\kappa^{(v)}$ and the mean-field $\widetilde{\mathbf{H}}_\kappa^{(v,w)}$ for this degree of freedom. The latter two have elements given according to

$$\rho^{(v)}_{\kappa, j_\kappa k_\kappa} = \langle \Phi^{(v)}_{j_\kappa} | \Phi^{(w)}_{k_\kappa} \rangle \delta_{vw} \tag{4.10}$$

$$\widetilde{H}^{(v,w)}_{\kappa, j_\kappa k_\kappa} = \langle \Phi^{(v)}_{j_\kappa} | \hat{H}^{(v,w)} | \Phi^{(w)}_{k_\kappa} \rangle \tag{4.11}$$

where we have introduced the single-hole functions $\Phi^{(v)}_{j_\kappa}$. The single-hole function $\Phi^{(v)}_{j_\kappa}$ is similar to the linear combination of Hartree products of the wavefunction ansatz in Eq. (4.5) except that it does not contain the single-particle functions for the κth degree of freedom, and the κth index in the coefficients is set to j. The mean-fields are thus operators on the κth degree of freedom.

4.1.2 Applicational Aspects of MCTDH

In MCTDH, the completely general and variationally optimized single-particle functions have to be represented in a time-independent basis which is normally achieved by the use of a DVR. For a given degree of freedom, such a DVR consists of a number of grid points. When the number of single-particle functions for a given degree of freedom equals the number of grid points, the single-particle function basis is complete. In this case, the single-particle functions become time-independent as the projection operator in Eq. (4.9) equals the identity, and one is left with a numerically exact standard grid-based wavepacket propagation approach.

The MCTDH equations of motion in principle involve N- and $(N-1)$-dimensional integrals in the Hamiltonian and mean-field matrix elements respectively. Evaluation of such integrals severely limits the applicability of MCTDH. Therefore, the Hamiltonian is usually restricted to a sum of products of functions only operating on one degree of freedom

$$\hat{H} = \sum_j c_j \prod_\kappa^N \hat{h}_{j_\kappa} \tag{4.12}$$

Using this product representation of the Hamiltonian, only one-dimensional integrals have to be evaluated. Another option is to use a time-dependent DVR with corrections, the so-called correlation DVR, but this approach will not be discussed further, and the reader is referred to Refs. [29] and [30] for further details.

The restriction on the form of the Hamiltonian in Eq. 4.12 implies that one has to chose an appropriate coordinate system. One choice is rectilinear Cartesian coordinates, however, these are often not the most chemically and physically intuitive set of coordinates. In Cartesian coordinates, the kinetic energy operator is diagonal in the nuclear degrees of freedom and, thus, of the form of Eq. 4.12. However, a representation of the potential energy operator which also conforms to the product form is needed. This can be obtained by use of the POTFIT program, but for higher dimensional systems this can be a restrictive step [31–33]. Another choice of coordinate system, which will be explored in Sect. 4.3, is rectilinear dimensionless normal coordinates where the kinetic energy operator is also diagonal. These coordinates are employed in the VCHAM in which the potential is given by a product form as a Taylor series expansion [34–36]. Recently, the prospect of using general curvilinear polyspherical coordinates [37, 38] has been introduced [39] along with an automatic procedure for generating analytical kinetic energy operators of the form given in Eq. 4.12 [40]. The construction of the kinetic energy operator has previously been somewhat of a bottleneck as it was approached separately for each specific problem.

4.1.3 Parameterized Basis Functions in MCTDH

The MCTDH method as presented above is very flexible due to the description in terms of general single-particle functions and is able to treat systems much larger than grid-based wavepacket methods. To treat even larger and more complex systems, approximations can be introduced by restricting the single-particle functions of some degrees of freedom to parameterized basis functions constrained to a Gaussian functional form [21, 22, 41]. This approximation yields the method termed Gaussian MCTDH (G-MCTDH), [41] and in the limit that all degrees of freedom use parameterized functions the vMCG method [21, 22]. As the wavefunction in the latter is completely described in terms of parameterized functions, the need for an underlying DVR is obviated. We will briefly present vMCG, and the reader is referred to Refs. [22] and [41] for an in-depth description. In vMCG, the form of the wavefunction is similar to that of MCTDH given in Eq. 4.5, however, the single-particle functions are now multi-dimensional complex Gaussians

$$\Phi^{(v)}(\mathbf{R}, t) = \sum_{j}^{n^{(v)}} C_j^{(v)}(t) \phi_j^{(v)}(\mathbf{R}; \zeta_j(t), \xi_j(t), \eta_j(t))$$

$$= \sum_{j}^{n^{(v)}} C_j^{(v)}(t) \exp\left[\mathbf{R}^{\mathrm{T}} \zeta_j(t)\mathbf{R} + \mathbf{R}^{\mathrm{T}} \xi_j(t) + \eta_j(t)\right] \tag{4.13}$$

Here, we have changed from the general coordinates $\mathbf{Q}$ to Cartesian coordinates $\mathbf{R}$. As opposed to MCTDH, j is a single, not a composite, index. Application of the Dirac-Frenkel variational principle again yields the equations of motion. The working equation for the coefficients reads

$$\mathbf{S}\dot{\mathbf{C}} = -i\left(\mathbf{H} - i\dot{\mathbf{S}}\right)\mathbf{C} \tag{4.14}$$

in terms of the vector of coefficients $\mathbf{C}$ and its time-derivative $\dot{\mathbf{C}}$. The elements of the overlap, right-acting time derivative, and Hamiltonian matrices are given by

$$S_{jk}^{(v,w)} = \langle \phi_j^{(v)} | \phi_k^{(w)} \rangle \delta_{vw} \tag{4.15}$$

$$\dot{S}_{jk}^{(v,w)} = \left\langle \phi_j^{(v)} \left| \frac{\partial}{\partial t} \phi_k^{(w)} \right. \right\rangle \delta_{vw} \tag{4.16}$$

$$H_{jk}^{(v,w)} = \langle \phi_j^{(v)} | \hat{H}^{(v,w)} | \phi_k^{(w)} \rangle \tag{4.17}$$

The equation for the coefficients is very similar to Eq. 4.7 of MCTDH, but unlike the single-particle functions of MCTDH, the Gaussians of vMCG are not orthogonal. This introduces the extra complexity of having to deal with the overlap matrix. The working equation for the Gaussian parameters is more involved

$$\dot{\boldsymbol{\Lambda}} = -i \mathbf{A}^{-1} \mathbf{Y} \tag{4.18}$$

The complex parameters are collected in a single vector $\Lambda_j = \{\eta_j, \boldsymbol{\xi}_j, \boldsymbol{\zeta}_j\}$ with components labeled by $\lambda_{j\alpha}$. The index $j\alpha$ refers to the αth parameter of the jth Gaussian function. The elements of the matrix $\mathbf{A}$ and the vector $\mathbf{Y}$ are given according to

$$A_{j\alpha,k\beta}^{(v)} = C_j^{(v)*} C_k^{(v)} \left(S_{\alpha\beta,jk}^{(v,v)} - \left[\mathbf{S}_{\alpha 0}^{(v,v)} \left(\mathbf{S}^{(v,v)} \right)^{-1} \mathbf{S}_{0\beta}^{(v,v)} \right]_{jk} \right) \tag{4.19}$$

$$Y_{j\alpha}^{(v)} = \sum_w^{N_e} \sum_k^{n^{(w)}} C_j^{(v)*} C_k^{(w)} \left(H_{\alpha 0,jk}^{(v,w)} - \left[\mathbf{S}_{\alpha 0}^{(v,v)} \left(\mathbf{S}^{(v,v)} \right)^{-1} \mathbf{H}^{(v,w)} \right]_{jk} \right) \tag{4.20}$$

The indices α and β refer to derivatives with respect to the Gaussian parameters whereas no derivative is taken when the index is 0

$$S_{\alpha\beta,jk}^{(v,w)} = \left\langle \frac{\partial \phi_j^{(v)}}{\partial \lambda_{j\alpha}} \left| \frac{\partial \phi_k^{(w)}}{\partial \lambda_{k\beta}} \right. \right\rangle \delta_{vw} \tag{4.21}$$

$$S_{\alpha 0,jk}^{(v,w)} = \left\langle \frac{\partial \phi_j^{(v)}}{\partial \lambda_{j\alpha}} \left| \phi_k^{(w)} \right. \right\rangle \delta_{vw} \tag{4.22}$$

$$H_{\alpha 0,jk}^{(v,w)} = \left\langle \frac{\partial \phi_j^{(v)}}{\partial \lambda_{j\alpha}} \left| \hat{H}^{(v,w)} \right| \phi_k^{(w)} \right\rangle \tag{4.23}$$

As can be gathered from the above equations, the parameters of the Gaussian functions are all coupled to each other as well as the wavefunction coefficients. This is a result of using the Dirac-Frenkel variational principle and entails that the Gaussian basis functions follow "quantum" trajectories. Therefore, vMCG can straightforwardly describe quantum effects such as tunneling and electronic state transfer as opposed to the quasi-classical FMS approach which handles this through the spawning procedure. FMS will be outlined in the next section.

4.2 Full Multiple Spawning and Ab Initio Multiple Spawning

The field of Gaussian wavepacket dynamics was pioneered by Heller in a series of seminal papers [9, 10, 42] and has inspired many approximate methods of quantum dynamics since. FMS and AIMS can be viewed as simple extensions of the semi-classical Gaussian wavepacket prescription of Heller to include quantum mechanical aspects such as coupling of the Gaussian amplitudes as well as a method to describe non-adiabatic effects. FMS and AIMS can also be viewed as quasi-classical approximations to the full quantum dynamical vMCG method (although they preceded vMCG), and the similarities and differences between these methods will be pointed out in what follows. It should be noted that the formalisms of FMS and AIMS are identical except for the method used to evaluate the potential energy, and we will refer to these methods collectively as FMS unless it is necessary to distinguish between the two. This section is partly based on Ref. VI.

4.2.1 The FMS Equations of Motion

In FMS, the total wavefunction is expanded in a sum of products of electronic and nuclear wavefunctions similar to the Born representation of Eq. (1.3) but truncated to N_e electronic states [11–14]

$$\Psi(\mathbf{r}, \mathbf{R}, t) = \sum_{v}^{N_e} \Phi^{(v)}(\mathbf{R}, t)\psi^{(v)}(\mathbf{r}; \mathbf{R}) \tag{4.24}$$

Here, $\Phi^{(v)}(\mathbf{R}, t)$ is the time-dependent nuclear wavefunction associated with electronic state v, $\psi^{(v)}(\mathbf{r}; \mathbf{R})$ is the electronic wavefunction of state v, and the Cartesian electronic and nuclear coordinates are referred to as $\mathbf{r}$ and $\mathbf{R}$ respectively. Usually, the adiabatic representation is used for the electronic functions in particular when the method is applied as its on-the-fly counterpart AIMS. However, the methodology described below is completely general and can be used with either the diabatic or adiabatic representation depending on convenience. The nuclear wavefunction is given as a superposition of multi-dimensional frozen Gaussians so-called trajectory basis functions (TBFs)

$$\Phi^{(v)}(\mathbf{R}, t) = \sum_{j}^{n^{(v)}(t)} C_j^{(v)}(t)\phi_j^{(v)}(\mathbf{R}; \mathbf{R}_j(t), \mathbf{P}_j(t), \gamma_j(t), \alpha) \tag{4.25}$$

Here, $n^{(v)}(t)$ is the number of TBFs associated with electronic state v, $C_j^{(v)}(t)$ is the complex amplitude for the jth TBF on electronic state v, and $\mathbf{R}_j(t)$, $\mathbf{P}_j(t)$,

$\gamma_j(t)$, and α is the center position, momentum, semi-classical phase, and width that parameterize the given TBF. Each multidimensional TBF is given as a product of one-dimensional frozen Gaussians over the N nuclear degrees of freedom labeled by κ

$$\phi_j^{(v)}(\mathbf{R}; \mathbf{R}_j(t), \mathbf{P}_j(t), \gamma_j(t), \boldsymbol{\alpha}) = e^{i\gamma_j(t)} \prod_{\kappa}^{N} \varphi_{j_\kappa}^{(v)}(R_\kappa; R_{j_\kappa}(t), P_{j_\kappa}(t), \alpha_\kappa) \quad (4.26)$$

with

$$\varphi_{j_\kappa}^{(v)}(R_\kappa; R_{j_\kappa}(t), P_{j_\kappa}(t), \alpha_\kappa) =$$
$$\left(\frac{2\alpha_\kappa}{\pi}\right)^{\frac{1}{4}} \exp\left[-\alpha_\kappa \left(R_\kappa - R_{j_\kappa}(t)\right)^2 + iP_{j_\kappa}(t) \left(R_\kappa - R_{j_\kappa}(t)\right)\right] \quad (4.27)$$

The index j_κ refers to the κth degree of freedom of the jth TBF. In a harmonic potential, the expectation values of the position and momentum of a Gaussian wavepacket, e.g. a TBF, will undergo classical equations of motion. This led Heller to the following classical equations of motion for the position and momenta parameters [9, 42]

$$\frac{\partial R_{j_\kappa}}{\partial t} = \frac{P_{j_\kappa}}{m_\kappa} \quad (4.28)$$

$$\frac{\partial P_{j_\kappa}}{\partial t} = -\frac{\partial V(\mathbf{R})}{\partial R_\kappa}\bigg|_{\mathbf{R}_j(t)} \quad (4.29)$$

where m_κ is the mass associated with nuclear degree of freedom κ and $V(\mathbf{R})$ is the potential energy experienced by the nuclei. The phase can be chosen to evolve according to a semi-classical prescription [11, 12, 42]

$$\frac{\partial \gamma_j}{\partial t} = -V(\mathbf{R}_j(t)) + \sum_{\kappa}^{N} \frac{\left(\left(P_{j_\kappa}(t)\right)^2 - 2\alpha_\kappa\right)}{2m_\kappa} \quad (4.30)$$

Without loss of generality, $\gamma_j(t)$ can be set equal to zero thereby effectively absorbing the oscillating phase into the complex coefficients $C_j^{(v)}(t)$. Inserting the anzats for the wavefunction from Eq. (4.24) into the time-dependent Schrödinger equation yields the equation of motion for the time-dependent complex coefficients

$$\mathbf{S}\dot{\mathbf{C}} = -i\left(\mathbf{H} - i\dot{\mathbf{S}}\right)\mathbf{C} \quad (4.31)$$

in terms of the vector of coefficients $\mathbf{C}$ and its time-derivative $\dot{\mathbf{C}}$. This equation is identical to Eq. (4.14) of vMCG with the same definitions for the matrix elements as given in Eqs. (4.15)–(4.17).

As observed, the formulation of FMS is closely related to that of vMCG. In fact, the TBFs of FMS are identical to the Gaussian functions of vMCG except for a normalization factor, if one makes the identifications

$$\zeta_{j_\kappa} = -\alpha_\kappa \tag{4.32}$$

$$\xi_{j_\kappa} = 2\alpha_\kappa R_{j_\kappa} + i P_{j_\kappa} \tag{4.33}$$

$$\eta_j = i\gamma_j - \sum_{\kappa}^{N} \left(\alpha_\kappa R_{j_\kappa}^2 + i P_{j_\kappa} R_{j_\kappa} \right) \tag{4.34}$$

The first equation entails freezing the width parameter of the Gaussian functions in vMCG similar to FMS. As opposed to vMCG, the parameters of the TBFs in FMS are uncoupled from those of the other TBFs and from the wavefunction coefficients. In FMS, each TBF follows a classical trajectory as opposed to the "quantum" trajectories of vMCG. The propagation of the coupled coefficients sets FMS (and vMCG) aside from semi-classical frozen Gaussian wavepacket methods such as those of Heller [42]. What sets FMS aside from both vMCG and the semi-classical methods is the adaptability of the basis set size, and, thus, the size of the vectors and matrices of Eq. (4.31), through the spawning procedure.

4.2.2 The Spawning Approach to an Adaptive Basis Set

The spawning procedure is a general method for adapting the size of the nuclear basis set to describe inherently quantum mechanical effects and has been applied to tunneling [43] as well as photoexcitation and non-adiabatic transitions [11, 12, 44]. We will present the spawning method in the context of electronic state transfer as this was employed in the calculations presented in Chap. 8.

In the spawning approach, the effectiveness of the basis set is constantly monitored, and new basis functions are spawned if needed. This is achieved by calculating an effective coupling between the electronic states included in the simulation at the center of the TBFs at each timestep [16, 45]. In an adiabatic electronic basis, this effective coupling is obtained by evaluating

$$\Lambda_{\text{eff}}^{(v,w)}(\mathbf{R}) = \left| \mathbf{d}^{(v,w)}(\mathbf{R}) \right| \tag{4.35}$$

whereas in a diabatic basis, the effective coupling is given according to

$$W_{\text{eff}}^{(v,w)}(\mathbf{R}) = \left| \frac{W^{(v,w)}(\mathbf{R})}{W^{(w,w)}(\mathbf{R}) - W^{(v,v)}(\mathbf{R})} \right| \tag{4.36}$$

By calculating the effective coupling, one is essentially evaluating the probability of population transfer had TBFs been present on the other states. If the effective

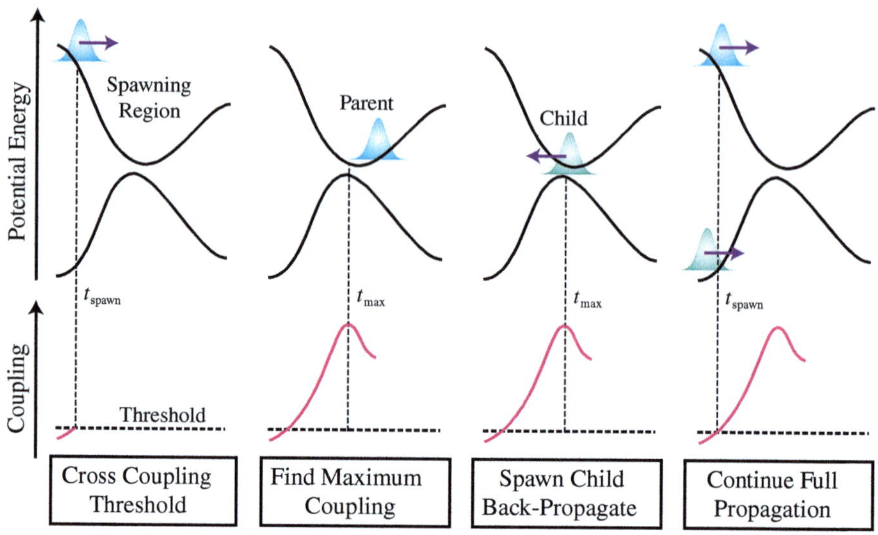

Fig. 4.1 Schematic showing the steps of the spawning procedure as employed in FMS and AIMS

coupling is above a certain pre-specified threshold, a new TBF, the child, is therefore spawned on the state to which a given TBF, the parent, is coupled. What remains is to position the child TBF in phase space. We will in the following restrict the discussion to an adiabatic electronic basis.

The first time at which the effective coupling crosses the threshold is referred to as t_{spawn}, cf. Fig. 4.1. At this timestep, the wavepacket is frozen and a copy of the parent is propagated forward in time into the spawning region until the effective coupling has peaked. The time at which the coupling peaks is referred to as t_{max}. With this knowledge at hand, three options have been used for placing the child TBF: p-jump, standard, and optimal spawning [45]. In the first approach, the child TBF is placed such that $\mathbf{R}_{\text{child}}$ matches $\mathbf{R}_{\text{parent}}$ at t_{max}, and $\mathbf{P}_{\text{child}}$ is scaled along the direction of the non-adiabatic coupling vector $\mathbf{d}$ to ensure energy conservation (in the long time limit) very similar to surface hopping [46]. However, situations where it is not possible to match the energy of the child to that of the parent by this approach can be encountered. This deficiency is amended in standard spawning where $\mathbf{R}_{\text{child}}$ is adjusted along the energy gradient of the child electronic state until the energy of the child matches that of the parent. In the optimal spawning approach, the child basis function is placed such as to maximize population transfer thereby minimizing the number of TBFs needed to describe non-adiabatic transitions. This is achieved by maximizing the norm of the electronic off-diagonal element of the Hamiltonian matrix coupling the parent and the child with $\mathbf{R}_{\text{child}}$ and $\mathbf{P}_{\text{child}}$ as parameters while constraining the energy of the child to match that of the parent.

When the child position and momentum have been determined, the child is back-propagated from t_{max} to t_{spawn} and added to the wavepacket by setting its coefficient $C_{\text{child}}(t_{\text{spawn}}) = 0$. It should be clear that at this point the child as a basis function only affords the opportunity to describe population transfer. The spawning procedure does not govern population transfer—population transfer is governed by the time-dependent Schrödinger equation for the coefficients given in Eq. (4.31).

4.2.3 Applicational Aspects of FMS and AIMS

A key aspect of Gaussian wavepacket dynamics is that the Gaussian functions are localized in space. Due to this property, only local information of the potential energy surfaces around the centers of the TBFs or centroids of overlapping TBFs is needed. Usually, the potential energy surfaces are expanded in a Taylor series according to

$$V(\mathbf{R}) = V(\mathbf{R}_0) + \left. \frac{\partial V(\mathbf{R})}{\partial \mathbf{R}} \right|_{\mathbf{R}_0} \cdot (\mathbf{R} - \mathbf{R}_0)$$
$$+ \frac{1}{2}(\mathbf{R} - \mathbf{R}_0)^{\mathrm{T}} \cdot \left. \frac{\partial^2 V(\mathbf{R})}{\partial \mathbf{R}^2} \right|_{\mathbf{R}_0} \cdot (\mathbf{R} - \mathbf{R}_0) + \cdots \qquad (4.37)$$

Truncating the series at second-order leads to the local harmonic approximation—a choice which is motivated by Heller's observation that the center of a Gaussian wavepacket follows a classical trajectory in a harmonic potential [42]. With an approximation to the potential in the form of Eq. (4.37), all matrix elements required for FMS as well as vMCG have analytical expressions in terms of Gaussian moments. These features of Gaussian wavepacket dynamics allow for use in direct dynamics methods [47] such as Direct Dynamics vMCG (DD-vMCG) [33, 48–50] and AIMS. In direct dynamics methods, the potential energy surfaces are calculated on-the-fly, i.e. simultaneously with the solution of the equations for the nuclear wavefunction. One advantage of FMS over vMCG is the direct applicability of the former in the adiabatic representation whereby the output from electronic structure codes can be used directly. In contrast, vMCG is only applicable in the diabatic representation necessitating that the adiabatic potential energy surfaces obtained from electronic structure calculations are transformed to the diabatic representation.

In nuclear dynamics simulations, the electronic structure calculations can be the limiting factor in terms of computational effort. As a consequence, the expansion in Eq. (4.37) is often truncated at zeroth-order in FMS leading to the following matrix elements of the potential energy operator

$$\langle \phi_j^{(v)} | \hat{V} | \phi_k^{(v)} \rangle \approx S_{jk}^{(v,v)} V(\mathbf{R}_c) \qquad (4.38)$$

Here, $\mathbf{R}_c$ is the centroid of the two TBFs given by $\mathbf{R}_c = (\mathbf{R}_j + \mathbf{R}_k)/2$ which for $j = k$ reduces to the center of the TBF. In the case of strongly coupled TBFs on the same electronic state, i.e. highly overlapping, this approximation can lead to detrimental results as observed in the example in Fig. 4.2. However, the independent first generation (IFG) approximation is usually employed [51]. In IFG, the initial

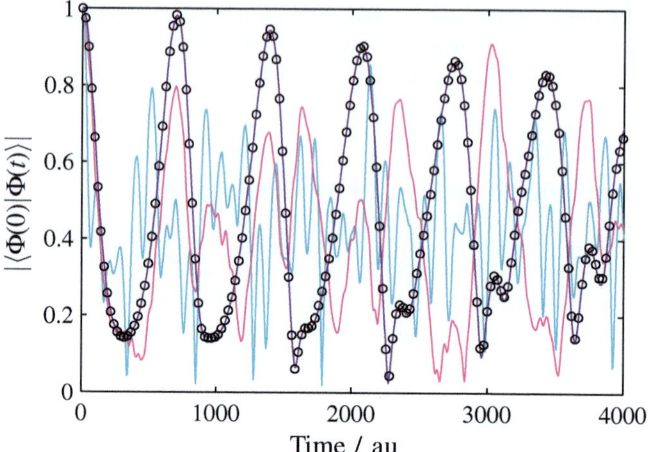

Fig. 4.2 Autocorrelation function for a wavepacket in a Morse potential. FMS results using a zeroth-(blue), first- (pink), and second-order (purple) Taylor expansion of the potential is compared to the exact result from an MCTDH calculation (○). The FMS calculations use 24 TBFs distributed on the three lowest classical orbits in phase space propagated using a fourth-order McLachlan Atela integrator for the classical parameters combined with a fourth-order Magnus expansion for the coefficients. A timestep of 5.0 au was used

TBFs are uncoupled and only couple to the TBFs that they spawn onto other electronic states whereby this problem is often circumvented.

Figure 4.2 also illustrates how FMS can be converged to the exact numerical result even for non-harmonic potentials. As opposed to the case for MCTDH where the single-particle functions become time-independent in the complete basis set limit ensuring the convergence to the exact numerical result obtained from grid-based methods, this is not the case for FMS. Consequently, a basis set that is complete at one time is not necessarily complete at a later time, and convergence to the exact numerical result is, thus, only obtained in the limit of a continuously complete basis set. Although this limit can somewhat easily be reached in low-dimensional bound systems, a way of approaching this limit for general high-dimensional cases is not straightforward. Reaching convergence would require an at times significantly over-complete basis set to ensure completeness at all subsequent times.

Many different integrators have been implemented into FMS to integrate the classical parameters and the coefficient vector. In one scheme, the classical variables are integrated from $t \rightarrow t + \delta t$ using a Velocity-Verlet integrator [52]

$$R_{j_\kappa}(t + \delta t) = R_{j_\kappa}(t) + \frac{P_{j_\kappa}(t)}{m_\kappa} \delta t - \frac{1}{2m_\kappa} \left(\frac{\partial V(\mathbf{R})}{\partial R_\kappa} \bigg|_{\mathbf{R}_j(t)} \right) \delta t^2 \tag{4.39}$$

$$P_{j_\kappa}(t + \delta t) = P_{j_\kappa}(t) - \frac{1}{2}\left(\left.\frac{\partial V(\mathbf{R})}{\partial R_\kappa}\right|_{\mathbf{R}_j(t)} + \left.\frac{\partial V(\mathbf{R})}{\partial R_\kappa}\right|_{\mathbf{R}_j(t+\delta t)} \right)\delta t \tag{4.40}$$

and the phase is integrated (if employed) using a simple trapezoidal rule [53]

$$\gamma_j(t + \delta t) = \gamma_j(t) + \frac{1}{2}\left(\left.\frac{\partial \gamma_j}{\partial t}\right|_t + \left.\frac{\partial \gamma_j}{\partial t}\right|_{t+\delta t} \right)\delta t \tag{4.41}$$

To integrate the coefficients, the differential equation for the complex coefficients is rewritten to

$$\dot{\mathbf{C}} = -i\mathbf{S}^{-1}\left(\mathbf{H} - i\dot{\mathbf{S}}\right)\mathbf{C} = \mathbf{B}(t)\mathbf{C} \tag{4.42}$$

The classical timestep is split into two, and for each substep, the integration of the coefficients is divided into $2n$ second-order Runge-Kutta integration steps. For $k = 0, \ldots, n - 1$, the following equations [53]

$$\mathbf{k}_{1,k} = \frac{\delta t}{2n}\mathbf{B}(t)\mathbf{C}\left(t + \frac{k}{2n}\delta t\right) \tag{4.43}$$

$$\mathbf{k}_{2,k} = \frac{\delta t}{2n}\mathbf{B}(t)\left(\mathbf{C}\left(t + \frac{k}{2n}\delta t\right) + \frac{1}{2}\mathbf{k}_{1,k}\right) \tag{4.44}$$

$$\mathbf{C}\left(t + \frac{k+1}{2n}\delta t\right) = \mathbf{C}\left(t + \frac{k}{2n}\delta t\right) + \mathbf{k}_{2,k} \tag{4.45}$$

are used to propagate the coefficients from $t \to t + \delta t/2$. In a similar manner, the equations

$$\mathbf{k}_{1,k+n} = \frac{\delta t}{2n}\mathbf{B}(t + \delta t)\mathbf{C}\left(t + \frac{k+n}{2n}\delta t\right) \tag{4.46}$$

$$\mathbf{k}_{2,k+n} = \frac{\delta t}{2n}\mathbf{B}(t + \delta t)\left(\mathbf{C}\left(t + \frac{k+n}{2n}\delta t\right) + \frac{1}{2}\mathbf{k}_{1,k+n}\right) \tag{4.47}$$

$$\mathbf{C}\left(t + \frac{k+n+1}{2n}\delta t\right) = \mathbf{C}\left(t + \frac{k+n}{2n}\delta t\right) + \mathbf{k}_{2,k+n} \tag{4.48}$$

are used for propagating the coefficients from $t + \delta t/2 \to t + \delta t$. The value of n is increased until convergence in the norm of the wavefunction is reached.

In another scheme, the classical parameters are propagated from $t \to t + \delta t/2$ and then subsequently from $t + \delta t/2 \to t + \delta t$ using a fourth-order McLachlan Atela integrator [54]. The integration is performed for $k = 1, \ldots, 4$ according to the recursion relations

$$P_{j_\kappa}^{(k)} = P_{j_\kappa}^{(k-1)} - b^{(k)}\frac{\delta t}{2}\left.\frac{\partial V(\mathbf{R})}{\partial R_\kappa}\right|_{R_{j_\kappa}^{(k-1)}} \tag{4.49}$$

$$R_{j_\kappa}^{(k)} = R_{j_\kappa}^{(k-1)} + a^{(k)} \frac{\delta t}{2} \frac{P_{j_\kappa}^{(k)}}{m_\kappa} \tag{4.50}$$

with the parameters for $a^{(k)}$ and $b^{(k)}$ from the si4.b integrator in Ref. [55]. In this way, $\mathbf{B}(t)$ can be explicitly constructed at times t, $t + \delta t/2$, and $t + \delta t$ and used for the integration of the coefficients. Using the Magnus expansion [56], the formal solution to Eq. (4.42) for the coefficients is

$$\mathbf{C}(t + \delta t) = \exp\left[\mathbf{\Omega}(t + \delta t, t)\right] \mathbf{C}(t) \tag{4.51}$$

with

$$\mathbf{\Omega}(t + \delta t, t) = \sum_k^\infty \mathbf{\Omega}_k(t + \delta t, t) \tag{4.52}$$

Each term $\mathbf{\Omega}_k$ in the infinite sum is a multiple integral involving linear combinations of nested commutators of $\mathbf{B}(t)$. Thus, the formal solution requires the complete knowledge of the time-dependence of $\mathbf{B}(t)$ as well as the evaluation of an infinite number of integrals in the exponential. In applications, the series is truncated to some appropriate order. In one approximation scheme, the integrals in the truncated series are furthermore approximated by quadratures using derivatives of $\mathbf{B}(t)$ around $t_{1/2} = t + \delta t/2$ [57]. This leads to the following solutions for the Magnus propagator of order 2 and 4 in the timestep respectively

$$\exp\left[\mathbf{\Omega}(t + \delta t, t)\right] \approx \exp\left[\delta t \mathbf{b}_0 + O(\delta t^3)\right] \tag{4.53}$$

$$\exp\left[\mathbf{\Omega}(t + \delta t, t)\right] \approx \exp\left[\delta t \mathbf{b}_0 + \frac{\delta t^3}{12}(\mathbf{b}_2 - [\mathbf{b}_0, \mathbf{b}_1]) + O(\delta t^5)\right] \tag{4.54}$$

where $\mathbf{b}_k$ is the kth derivative of $\mathbf{B}(t)$ at time $t = t_{1/2}$. These derivatives are evaluated by finite differences

$$\mathbf{b}_0 = \mathbf{B}(t + \delta t/2) \approx \frac{\mathbf{B}(t) + \mathbf{B}(t + \delta t)}{2} \tag{4.55}$$

$$\mathbf{b}_1 = \frac{\mathbf{B}(t + \delta t) - \mathbf{B}(t)}{\delta t} \tag{4.56}$$

$$\mathbf{b}_2 = \frac{\mathbf{B}(t + \delta t) - 2\mathbf{B}(t + \delta t/2) + \mathbf{B}(t)}{(\delta t/2)^2} \tag{4.57}$$

The fourth-order Magnus propagator is combined with the fourth-order McLachlan Atela integrator. The second-order Magnus propagator can be combined with the Velocity-Verlet integrator for the classical parameters either by approximating $\mathbf{B}(t + \delta t/2)$ as indicated in Eq. (4.55) or dividing the classical timestep into two and explicitly evaluating $\mathbf{B}(t + \delta t/2)$.

4.3 The Vibronic Coupling Hamiltonian

As mentioned in Sect. 4.1.2, one has to chose an appropriate coordinate system to ensure that the Hamiltonian for MCTDH is in the product form of Eq. (4.12). This can be achieved by use of dimensionless, i.e. mass-frequency scaled, normal coordinates as employed for the VCHAM [34–36]. The VCHAM has been used in conjunction with MCTDH several times [58–65]. In the vibronic coupling scheme, the Hamiltonian is represented in a basis of diabatic electronic states as an expansion around a point of interest $\mathbf{Q}_0$ normally chosen as the ground state equilibrium geometry. A system of N_e electronic states is described by an $N_e \times N_e$ matrix which is decomposed into a zeroth-order Hamiltonian $\mathbf{H}_0 = \mathbf{V}_0 + \mathbf{T}$ and a diabatic potential coupling matrix $\mathbf{W}$

$$\mathbf{H} = \mathbf{V}_0 + \mathbf{T} + \mathbf{W} = \sum_{\kappa}^{N} \frac{\omega_\kappa}{2} \left(Q_\kappa^2 - \frac{\partial^2}{\partial Q_\kappa^2} \right) \mathbf{I} + \mathbf{W} \tag{4.58}$$

Here, ω_κ is the ground state normal mode frequency, and $\mathbf{I}$ is the $N_e \times N_e$ identity matrix. The electronic on- and off-diagonal terms of $\mathbf{W}$ are expanded in Taylor series according to

$$W^{(v,v)} = E^{(v)} + \sum_{\kappa}^{N} \beta_\kappa^{(v)} Q_\kappa + \frac{1}{2} \sum_{\kappa}^{N} \sum_{\kappa'}^{N} \gamma_{\kappa\kappa'}^{(v)} Q_\kappa Q_{\kappa'} + \cdots \tag{4.59}$$

$$W^{(v,w)} = \sum_{\kappa}^{N} \lambda_\kappa^{(v,w)} Q_\kappa + \frac{1}{2} \sum_{\kappa}^{N} \sum_{\kappa'}^{N} \mu_{\kappa\kappa'}^{(v,w)} Q_\kappa Q_{\kappa'} + \cdots \tag{4.60}$$

The coefficients $\beta_\kappa^{(v)}$, $\mu_{\kappa\kappa'}^{(v,w)}$ etc. are called vibronic coupling constants. As the diabatic potential energy surfaces are smooth functions of the nuclear coordinates, even a low-order Taylor expansion should give a reasonable representation of the system at hand. In cases of large anharmonicity, it can beneficial to employ a Morse potential as the zeroth-order term instead of the harmonic term in Eq. (4.58)

$$V_\kappa^{(v)}(Q_\kappa) = D_\kappa^{(v)} \left(\exp\left[-\alpha_\kappa^{(v)}(Q_\kappa - Q_{\kappa 0}^{(v)}) \right] - 1 \right)^2 \tag{4.61}$$

Many of the coupling constants vanish on grounds of point group symmetry. As an example, take the coupling constant $\mu_{\kappa\kappa'}^{(v,w)}$. Let Γ_v and Γ_w be the irreducible representations of the electronic states v and w in the pertinent point group, and, similarly, let Γ_κ and $\Gamma_{\kappa'}$ be those of the normal coordinates Q_κ and $Q_{\kappa'}$. The coupling constant can only be non-vanishing if the following is fulfilled

$$\Gamma_v \otimes \Gamma_\kappa \otimes \Gamma_{\kappa'} \otimes \Gamma_w \supseteq \Gamma_A \tag{4.62}$$

where Γ_A is the totally symmetric, i.e. identity representation, of the point group.

4.3.1 Applicational Aspects of the VCHAM

The VCHAM provides a convenient representation of the Hamiltonian to be used in conjunction with MCTDH. However, to be employed, the coupling constants have to be determined in some manner. As the potential energy surfaces obtained from electronic structure calculations are in the adiabatic representation they cannot directly be employed as the transformation from an adiabatic to a diabatic representation is not unique. On the other hand, the transformation from the diabatic to the adiabatic representation is unique and is achieved by diagonalization of the full diabatic potential matrix, i.e. $\mathbf{V}_0 + \mathbf{W}$. This matrix is in general non-diagonal. Thus, with a suitable guess for the coupling constants, the potential matrix can be diagonalized, and the obtained adiabatic potential energy surfaces can be compared to those obtained from electronic structure calculations. This procedure gives a method of fitting the coupling constants. The general procedure to obtain the VCHAM can be given in four steps

(1) Calculate harmonic normal mode frequencies and vectors and transform from mass-weighted normal coordinates to dimensionless coordinates by $Q_\kappa \rightarrow Q_\kappa/\sqrt{\omega_\kappa}$
(2) Use the transformed normal mode vectors to create displaced geometries and calculate the energies of the desired electronic states at these geometries
(3) Create a database containing the geometries and energies
(4) Fit the coupling constants by comparing the energies at the displaced geometries in the database to those obtained by diagonalization of the diabatic potential matrix from the VCHAM

From the procedure detailed above, it could appear that one has to calculate the electronic energies on a full-dimensional grid of points. This would be a severe restriction when going beyond two or three dimensions. Instead, energies are calculated along single-mode displacements to fit the on-diagonal constants, i.e. those for which $\kappa = \kappa' = \cdots$, and along mode-mode diagonals to fit the off-diagonal constants, i.e. those for which $\kappa \neq \kappa' \neq \cdots$. If terms of higher than second order are included in the Taylor series of Eqs. 4.59 and 4.60, this procedure necessitates that these are restricted. For the third order terms, the following restrictions are made

$$\frac{1}{6}\sum_\kappa^N \sum_{\kappa'}^N \sum_{\kappa''}^N \iota_{\kappa\kappa'\kappa''}^{(v)} Q_\kappa Q_{\kappa'} Q_{\kappa''} \rightarrow \frac{1}{6}\sum_\kappa^N \iota_{\kappa\kappa}^{(v)} Q_\kappa^3 + \frac{1}{2}\sum_{\kappa\neq\kappa'}^N \iota_{\kappa\kappa'}^{(v)} Q_\kappa Q_{\kappa'}^2 \qquad (4.63)$$

$$\frac{1}{6}\sum_\kappa^N \sum_{\kappa'}^N \sum_{\kappa''}^N \eta_{\kappa\kappa'\kappa''}^{(v,w)} Q_\kappa Q_{\kappa'} Q_{\kappa''} \rightarrow \frac{1}{6}\sum_\kappa^N \eta_{\kappa\kappa}^{(v,w)} Q_\kappa^3 + \frac{1}{2}\sum_{\kappa\neq\kappa'}^N \eta_{\kappa\kappa'}^{(v,w)} Q_\kappa Q_{\kappa'}^2 \qquad (4.64)$$

If these restrictions were not employed, the constants $\iota_{\kappa\kappa'\kappa''}^{(v)}$ and $\eta_{\kappa\kappa'\kappa''}^{(v,w)}$ would in some cases be underdetermined.

Fitting the coupling constants is a general non-linear optimization problem, and, whence, suffers from general problems in terms of converging to the global minimum. Furthermore, since the coupling constants are highly interdependent, one is only certain of obtaining a set of constants but not necessarily the "correct" set of constants to describe the system. To alleviate some of these difficulties, the low-order constants are fitted first and then used as a guess for an optimization including higher-order terms. This ensures to some extent that the low-order constants carry the most weight in the fit. Furthermore, the data points can be exponentially weighted by energy to favor a good fit to the important low-energy regions of the potential energy surfaces. Nonetheless, even for a modest number of modes, unacceptable un-balanced fits are often obtained. Examples of this are first-order inter-state or third-order coupling constants giving rise to potential energy surfaces with unphysically low barriers to dissociation in one specific mode while all order modes are fitted well. In this case, it can be necessary to restrict, fix, or manually adjust parameters to achieve a more balanced fit.

The entire fitting procedure can be carried out using the VCHFIT program which is distributed with the Heidelberg MCTDH code [66]. VCHFIT takes advantage of the symmetry restraints of Eq. (4.62) and furthermore includes several different optimization routines from simplex and Powell's conjugate direction methods to a genetic algorithm [67–69]. The program was inhere employed in a locally modified version.

References

1. J.C. Light, I.P. Hamilton, J.P. Vill, J. Chem. Phys. **82**, 1400–1409 (1985)
2. Z. Bačić, J.C. Light, Annu. Rev. Phys. Chem. **40**, 469–498 (1989)
3. M.D. Feit, J.A. Fleck, A. Steiger, J. Comput. Phys. **47**, 412–433 (1982)
4. M.D. Feit, J.A. Fleck, J. Chem. Phys. **78**, 301–308 (1983)
5. D. Kosloff, R. Kosloff, J. Comput. Phys. **52**, 35–53 (1983)
6. R. Kosloff, J. Phys. Chem. **92**, 2087–2100 (1988)
7. E.J. Heller, Chem. Phys. Lett. **34**, 321–325 (1975)
8. K.C. Kulander, E.J. Heller, J. Chem. Phys. **69**, 2439–2449 (1978)
9. E.J. Heller, J. Chem. Phys. **75**, 2923–2931 (1981)
10. E.J. Heller, Acc. Chem. Res. **14**, 368–375 (1981)
11. T.J. Martínez, M. Ben-Nun, G. Ashkenazi, J. Chem. Phys. **104**, 2847–2856 (1996)
12. T.J. Martínez, M. Ben-Nun, R.D. Levine, J. Phys. Chem. **100**, 7884–7895 (1996)
13. T.J. Martínez, R.D. Levine, J. Chem. Soc. Faraday Trans. **93**, 941–947 (1997)
14. M. Ben-Nun, T.J. Martínez, J. Chem. Phys. **108**, 7244–7257 (1998)
15. M. Ben-Nun, J. Quenneville, T.J. Martínez, J. Phys. Chem. A **104**, 5161–5175 (2000)
16. M. Ben-Nun, T.J. Martínez, Adv. Chem. Phys. **121**, 439–512 (2002)
17. D.V. Shalashilin, M.S. Child, J. Chem. Phys. **113**, 10028–10036 (2000)
18. D.V. Shalashilin, M.S. Child, J. Chem. Phys. **114**, 9296–9304 (2001)
19. D.V. Shalashilin, M.S. Child, J. Chem. Phys. **115**, 5367–5375 (2001)
20. D.V. Shalashilin, M.S. Child, Chem. Phys. **304**, 103–120 (2004)
21. G. Worth, I. Burghardt, Chem. Phys. Lett. **368**, 502–508 (2003)
22. G.A. Worth, M.A. Robb, I. Burghardt, Faraday Discuss. **127**, 307–323 (2004)
23. H.-D. Meyer, U. Manthe, L.S. Cederbaum, Chem. Phys. Lett. **165**, 73–78 (1990)

24. M.H. Beck, A. Jäckle, G.A. Worth, H.-D. Meyer, Phys. Rep. **324**, 1–105 (2000)
25. H.-D. Meyer, G.A. Worth, Theor. Chem. Acc. **109**, 251–267 (2003)
26. G.A. Worth, H.-D. Meyer, H. Köppel, L.S. Cederbaum, I. Burghardt, Int. Rev. Phys. Chem. **27**, 569–606 (2008)
27. U. Manthe, H.-D. Meyer, L.S. Cederbaum, J. Chem. Phys. **97**, 3199–3213 (1992)
28. J. Frenkel, *Wave Mechanics—Advanced General Theory* (Oxford University Press, Oxford, 1934)
29. U. Manthe, J. Chem. Phys. **105**, 6989–6994 (1996)
30. R. van Harrevelt, U. Manthe, J. Chem. Phys. **123**, 064106 (2005)
31. A. Jäckle, H.-D. Meyer, J. Chem. Phys. **104**, 7974–7984 (1996)
32. A. Jäckle, H.-D. Meyer, J. Chem. Phys. **109**, 3772–3779 (1998)
33. H.-D. Meyer, F. Gatti, G. Worth (eds.), *Multidimensional Quantum Dynamics—MCTDH Theory and Applications* (Wiley-VCH, Weinheim, 2009)
34. L.S. Cederbaum, W. Domcke, H. Köppel, W. von Niessen, Chem. Phys. **26**, 169–177 (1977)
35. L.S. Cederbaum, H. Köppel, W. Domcke, Int. J. Quantum Chem. **20**, 251–267 (1981)
36. H. Köppel, W. Domcke, L.S. Cederbaum, Adv. Chem. Phys. **57**, 59–246 (1984)
37. X. Chapuisat, C. Iung, Phys. Rev. A **45**, 6217–6235 (1992)
38. F. Gatti, C. Iung, Phys. Rep. **484**, 1–69 (2009)
39. L. Joubert-Doriol, B. Lasorne, F. Gatti, M. Schröder, O. Vendrell, H.-D. Meyer, Comput. Theor. Chem. **990**, 75–89 (2012)
40. M. Ndong, L. Joubert-Doriol, H.-D. Meyer, A. Nauts, F. Gatti, D. Lauvergnat, J. Chem. Phys. **136**, 034107 (2012)
41. I. Burghardt, H.-D. Meyer, L.S. Cederbaum, J. Chem. Phys. **111**, 2927–2939 (1999)
42. E.J. Heller, J. Chem. Phys. **62**, 1544–1555 (1975)
43. M. Ben-Nun, T.J. Martínez, J. Chem. Phys. **112**, 6113–6121 (2000)
44. J. Quenneville, M. Ben-Nun, T.J. Martínez, J. Photochem. Photobiol. A **144**, 229–235 (2001)
45. S. Yang, J.D. Coe, B. Kaduk, T.J. Martínez, J. Chem. Phys. **130**, 134113 (2009)
46. J.C. Tully, J. Chem. Phys. **93**, 1061–1071 (1990)
47. G.A. Worth, M.A. Robb, B. Lasorne, Mol. Phys. **106**, 2077–2091 (2008)
48. B. Lasorne, M.J. Bearpark, M.A. Robb, G.A. Worth, Chem. Phys. Lett. **432**, 604–609 (2006)
49. B. Lasorne, M.A. Robb, G.A. Worth, Phys. Chem. Chem. Phys. **9**, 3210–3227 (2007)
50. B. Lasorne, M.J. Bearpark, M.A. Robb, G.A. Worth, J. Phys. Chem. A **112**, 13017–13027 (2008)
51. M.D. Hack, A.M. Wensmann, D.G. Truhlar, M. Ben-Nun, T.J. Martínez, J. Chem. Phys. **115**, 1172–1186 (2001)
52. W.C. Swope, H.C. Andersen, P.H. Berens, K.R. Wilson, J. Chem. Phys. **76**, 637–649 (1982)
53. B.G. Levine, J.D. Coe, A.M. Virshup, T.J. Martínez, Chem. Phys. **347**, 3–16 (2008)
54. R.I. McLachlan, P. Atela, Nonlinearity **5**, 541–562 (1992)
55. S.K. Gray, D.W. Noid, B.G. Sumpter, J. Chem. Phys. **101**, 4062–4072 (1994)
56. W. Magnus, Commun. Pure Appl. Math. **7**, 649–673 (1954)
57. S. Blanes, F. Casas, J. Ros, BIT **40**, 434–450 (2000)
58. G.A. Worth, H.-D. Meyer, L.S. Cederbaum, J. Chem. Phys. **105**, 4412–4426 (1996)
59. G.A. Worth, H.-D. Meyer, L.S. Cederbaum, J. Chem. Phys. **109**, 3518–3529 (1998)
60. G.A. Worth, H.-D. Meyer, L.S. Cederbaum, Chem. Phys. Lett. **299**, 451–456 (1999)
61. A. Raab, G.A. Worth, H.-D. Meyer, L.S. Cederbaum, J. Chem. Phys. **110**, 936–946 (1999)
62. S. Mahapatra, G.A. Worth, H.-D. Meyer, L.S. Cederbaum, H. Köppel, J. Phys. Chem. A **105**, 5567–5576 (2001)
63. T.S. Venkatesan, S. Mahapatra, H.-D. Meyer, H. Köppel, L.S. Cederbaum, J. Phys. Chem. A **111**, 1746–1761 (2007)
64. G.A. Worth, R.E. Carley, H.H. Fielding, Chem. Phys. **338**, 220–227 (2007)
65. T.J. Penfold, G.A. Worth, J. Chem. Phys. **131**, 064303 (2009)
66. G. A. Worth, M. H. Beck, A. Jäckle, H.-D. Meyer, The MCTDH Package, Version 8.2, (2000). H.-D. Meyer, Version 8.3 (2002), Version 8.4 (2007), see http://mctdh.uni-hd.de
67. J.A. Nelder, R. Mead, Comput. J. **7**, 308–313 (1965)
68. M.J.D. Powell, Comput. J. **7**, 155–162 (1964)
69. W.H. Press, S.A. Teukolsky, W.T. Vetterling, B.P. Flannery, *Numerical Recipes in FORTRAN—The Art of Scientific Computing*, 2nd edn. (Cambridge University Press, Cambridge, 1992)

Chapter 5
Time-Resolved Photoelectron Spectra

Through the time evolution of the wavefunction, dynamics simulations provide insight into the system under investigation. Often, it can be beneficial to consider partially integrated quantities such as electronic populations or coordinate expectation values to map out the dynamics. However, if we are interested in comparing simulations to experiment, it can be necessary to calculate the experimental observable from the dynamics simulations to allow for a direct comparison between theory and experiment.

In this chapter, we will describe two different methods of calculating time-resolved photo-electron spectra. Such calculations in principle involve propagation of the wavefunction in the presence of two fields representing the pump and probe pulses. In the first method, the initial excitation as well as the subsequent ionization are assumed to be impulsive, i.e. the duration of the pulses is taken to be shorter than that of nuclear dynamics. As the pulses are furthermore assumed non-overlapping, the ionization of the pump-induced wavepacket is treated by first order perturbation theory. In the second method, the two fields are explicitly included in the Hamiltonian of the system, however, the ionization step is approximated as the wavefunction of the photoelectron is not considered explicitly, and a unit transition strength is assumed.

5.1 Perturbative Trajectory-Based Calculation

The method presented inhere uses first order perturbation theory to calculate the photoelectron spectrum. As its basis, the method uses results obtained from trajectory-based dynamics such as Full Multiple Spawning (FMS) and is performed subsequent to the dynamics calculation. It is assumed that a wavepacket, approximated by a number of trajectories, has been created in a neutral excited state by the pump pulse, and that there is no overlap in time between the pump and probe pulses. I.e., the probe pulse strictly follows that of the pump. We will show how the method reduces to calculating the spectrum resulting from ionization of the pump-induced wavepacket

T. S. Kuhlman, *The Non-Ergodic Nature of Internal Conversion*, Springer Theses, DOI: 10.1007/978-3-319-00386-3_5, © Springer International Publishing Switzerland 2013

at a specific time $t = \Delta t$ following the pump pulse. Thus, the method is somewhat in the spirit of the Bersohn-Zewail model [1, 2] and follows the methods of Meier, Engel, and Møller [3–6].

The photoelectron spectrum can be calculated from the probability density of the electronic ionization continua long after the probe pulse has decayed, i.e. from the overlap between the state $|\Psi(t)\rangle$ for $t \to \infty$ with the electronic ionization continua [7, 8]

$$\sigma(E_k) = \sum_w |\langle \Psi(t \to \infty)|\psi^{(w)}(E_k)\rangle|^2 \tag{5.1}$$

Here, w labels the ionization continua, and E_k is the kinetic energy of the ejected photoelectron. The final electronic state w is a combination of an $(n - 1)$-electron cationic state and a continuum photoelectron. In the sudden and strong orthogonality approximations, the cation and photoelectron can be assumed independent, and the final electronic state reduces to a product of that of the cation and photoelectron [9–11]. In this case, the spectrum is given by

$$\sigma(E_k) = \sum_w \sum_f |\langle \Psi(t \to \infty)|\psi_f^{(el)}(E_k)\psi^{(w)}\rangle|^2 \tag{5.2}$$

where f labels the angular momentum of the photoelectron. In what follows, we will consider a single component corresponding to ionization to the cationic state w with the ejection of a photoelectron with angular momentum f. To calculate $|\Psi(t \to \infty)\rangle$, the initial state is taken to be a wavepacket in state v at a time after the action of the pump pulse but before the action of the probe pulse. We will in the end sum over all initial electronic states v to include the full wavepacket created by the pump pulse. Following first-order perturbation theory, the component of the initial state ending up in the final state w due to the perturbation of the probe field is [6]

$$\begin{aligned}
|\Psi(t \to \infty)\rangle &= -i \int_{-\infty}^{\infty} e^{i(\hat{H}^{(w)}+E_k)t} \hat{H}_I e^{-i\hat{H}^{(v)}t} |\Psi^{(v)}(t)\rangle \, dt \\
&= -i \int_{-\infty}^{\infty} e^{i(\hat{H}^{(w)}+E_k)t} \hat{H}_I e^{-i\hat{H}^{(v)}t} |\psi^{(v)}\rangle |\Phi^{(v)}(t)\rangle \, dt
\end{aligned} \tag{5.3}$$

where the perturbation is given by the light-matter interaction in the dipole approximation

$$\hat{H}_I = -\hat{\mu} \cdot \mathcal{E}(t) \tag{5.4}$$

Employing the rotating wave approximation, we represent the probe pulse by the field

$$\mathcal{E}(t) = \frac{1}{2}\varepsilon(t) \exp[-i\omega t] = \frac{1}{2}\varepsilon_0 \exp[-4\ln(2)(t - \Delta t)^2/\tau^2] \exp[-i\omega t] \tag{5.5}$$

The field consists of a carrier wave of frequency ω multiplied by a Gaussian field envelope $\varepsilon(t)$. The field envelope is centered at the pump-probe delay Δt with a full width at half maximum (FWHM) of τ and an amplitude of ε_0. Using these expressions, the overlap in Eq. (5.2) can be given as

$$\langle \psi_f^{(el)}(E_k)\psi^{(w)}|\Psi(t \to \infty)\rangle =$$
$$\frac{i}{2}\int_{-\infty}^{\infty} e^{i(\hat{V}^{(w)}+\hat{T}_n+E_k)t}\langle\psi_f^{(el)}(E_k)\psi^{(w)}|\hat{\mu}|\psi^{(v)}\rangle\varepsilon(t)e^{-i\omega t}e^{-i(\hat{V}^{(v)}+\hat{T}_n)t}|\Phi^{(v)}(t)\rangle\,\mathrm{d}t$$

$$(5.6)$$

where $\hat{V}^{(v)}$ is the potential energy operator for state v, and $\hat{T}_n$ is the nuclear kinetic energy operator. Due to the approximations for the electronic state of the ionization continuum, the energies of the cationic state and the photoelectron are additive. The electronic matrix element of the dipole operator can explicitly be written as

$$\mu_f^{(w,v)}(E_k) \equiv \langle\psi_f^{(el)}(E_k)\psi^{(w)}|\hat{\mu}|\psi^{(v)}\rangle = \langle\psi_f^{(el)}(E_k)|\hat{\mu}|\psi_D^{(w,v)}\rangle \qquad (5.7)$$

with the so-called Dyson-orbital given by [12–15]

$$\psi_D^{(w,v)}(\mathbf{r}_1) = \langle\mathbf{r}_1|\psi_D^{(w,v)}\rangle = \langle\psi^{(w)}|\langle\mathbf{r}_1|\psi^{(v)}\rangle$$
$$= \sqrt{n}\int_{\mathbf{r}_2}\cdots\int_{\mathbf{r}_n}\psi^{(w)}(\mathbf{r}_2,\ldots,\mathbf{r}_n)\psi^{(v)}(\mathbf{r}_1,\ldots,\mathbf{r}_n)\,\mathrm{d}\mathbf{r}_2\cdots\mathrm{d}\mathbf{r}_n \qquad (5.8)$$

Notice that the integration is over the coordinates of all n electrons except for those of the ejected photoelectron. The Dyson orbital is a one-electron ionization-channel specific orbital. Using a Coulomb radial function for the continuum electron, the electronic matrix element over the dipole operator in Eq. (5.7) can be calculated numerically using EZDYSON [16] when the Dyson orbital is available from electronic structure calculations. In this calculation, rotational averaging over the angle between the transition dipole moment and the electric field polarization is carried out.

To proceed with the evaluation of the integral in Eq. (5.6), we make the critical approximation that the potential energy, kinetic energy and dipole moment operators commute. Hereby, we can rewrite the integral to

$$\frac{i}{2}\mu_f^{(w,v)}(E_k)\int_{-\infty}^{\infty}\varepsilon(t)e^{-i(\hat{V}^{(v)}-\hat{V}^{(w)}-E_k+\omega)t}|\Phi^{(v)}(t)\rangle\,\mathrm{d}t =$$
$$\frac{i}{2}\sum_j\mu_f^{(w,v)}(E_k)\int_{-\infty}^{\infty}\varepsilon(t)e^{-i(\hat{V}^{(v)}-\hat{V}^{(w)}-E_k+\omega)t}C_j^{(v)}(t)|\phi_j^{(v)}(t)\rangle\,\mathrm{d}t \qquad (5.9)$$

In the second line, we have explicitly written the vibrational state as a sum over coherent states [17] following the prescription of FMS. In the nuclear coordinate representation, this becomes

$$\frac{i}{2} \sum_j \mu_f^{(w,v)}(E_k; \mathbf{R}) \times$$

$$\int_{-\infty}^{\infty} \varepsilon(t) e^{-i(V^{(v)}(\mathbf{R}) - V^{(w)}(\mathbf{R}) - E_k + \omega)t} C_j^{(v)}(t) \phi_j^{(v)}(\mathbf{R}, t) \, dt \qquad (5.10)$$

We have indicated the parametric dependence on the nuclear coordinates of the transition dipole matrix element which results from the parametric dependence on the nuclear coordinates of the electronic wavefunctions. To evaluate the expression in Eq. (5.2), we have to square the expression above and integrate over nuclear coordinates. As the trajectory basis function (TBF) $\phi_j^{(v)}(\mathbf{R}, t)$ is highly localized around $\mathbf{R} = \mathbf{R}_j(t)$, the coordinate integral can be approximated by evaluating the nuclear coordinate dependent functions in $\mathbf{R}_j(t)$. If we in addition assume impulsive ionization, i.e. that the field envelope $\varepsilon(t)$ is of much shorter duration than the timescale for nuclear motion, this reduces to evaluating the functions in $\mathbf{R} = \mathbf{R}_j(\Delta t)$ as the field envelope is centered at $t = \Delta t$. Thereby, we can also evaluate the coefficient and TBF in $t = \Delta t$ and move them outside the integral over t. Due to the localized nature of the TBFs, cross-terms between different TBFs in the absolute square, i.e. terms involving $\phi_k^{(v)*} \phi_j^{(v)}$ with $k \neq j$, are also omitted as the overlap between $\phi_k^{(v)}$ and $\phi_j^{(v)}$ is assumed negligible. Performing these operations, one component of the spectrum is given by

$$\frac{1}{4} \sum_j |\mu_f^{(w,v)}(E_k; \mathbf{R}_j(\Delta t))|^2 |C_j^{(v)}(\Delta t)|^2 \left| \int_{-\infty}^{\infty} \varepsilon(t) \exp\left[-i\widetilde{\omega}t\right] \, dt \right|^2 \qquad (5.11)$$

where we have defined the shifted frequency

$$\widetilde{\omega} = V^{(v)}(\mathbf{R}_j(\Delta t)) - V^{(w)}(\mathbf{R}_j(\Delta t)) - E_k + \omega$$
$$= \omega - \Delta E^{(v,w)}(\mathbf{R}_j(\Delta t)) - E_k \qquad (5.12)$$

The integral over the Gaussian field envelope can be solved analytically by identifying it as a Fourier transform in the shifted frequency

$$\mathcal{F}(\omega, E_k; \mathbf{R}_j(\Delta t)) = \left| \int_{-\infty}^{\infty} \varepsilon_0 \exp\left[-4\ln(2)(t - \Delta t)^2/\tau^2\right] \exp\left[-i\widetilde{\omega}t\right] \, dt \right|^2$$
$$= \left| \varepsilon_0 \frac{\tau}{2} \sqrt{\frac{\pi}{\ln(2)}} \exp\left[-i\widetilde{\omega}\Delta t\right] \exp\left[-\tau^2\widetilde{\omega}^2/(16\ln(2))\right] \right|^2$$
$$= \frac{\pi\varepsilon_0^2\tau^2}{4\ln(2)} \exp\left[-\tau^2\widetilde{\omega}^2/(8\ln(2))\right] \qquad (5.13)$$

With this at hand, the total spectrum at a specific time-delay between pump and probe pulses Δt is given by

$$\sigma(E_k, \Delta t) \propto$$

$$\sum_{v,w} \sum_f \sum_j |\mu_f^{(w,v)}(E_k; \mathbf{R}_j(\Delta t))|^2 |C_j^{(v)}(\Delta t)|^2 \mathcal{F}(\omega, E_k; \mathbf{R}_j(\Delta t)) \quad (5.14)$$

Here, we have summed over v to include all electronic components of the initial wavepacket. The formula for the spectrum in Eq. (5.14) consists of a product of the absolute square of the electronic transition dipole moment, a basis function population, and the so-called window function. The window function is peaked around $\omega = \Delta E^{(v,w)}(\mathbf{R}) + E_k$, i.e. when the probe frequency matches the sum of the energy difference between the cationic and the neutral state and the kinetic energy of the photoelectron. The energetic width of the window is determined by the FWHM of the probe pulse τ—a shorter pulse leads to a wider window in energy as expected from the time-bandwidth product. As derived, the spectrum is discrete in the pump-probe delay Δt. To approximately include the experimental time resolution given by the cross-correlation (XC) between the pump and probe pulses, the spectrum is convoluted by a Gaussian function in the time-delay Δt

$$\sigma(E_k, \Delta t) \rightarrow \exp[-4\ln(2)\Delta t^2/\tau_{xc}^2] \otimes \sigma(E_k, \Delta t) \quad (5.15)$$

Although a high-level correlated electronic structure method is used in the simulation which is the basis for the calculation of the time-resolved photoelectron spectrum, inevitably some discrepancy between the energies obtained from calculation and those from experiment is to be expected. In order to be able to compare calculated spectra to those from experiment, a correction can employed for $\Delta E^{(v,w)}$ to match to the experimental value of the photoelectron kinetic energy at the Franck-Condon point. The correction is achieved by $\Delta E^{(v,w)} \rightarrow \Delta E^{(v,w)} - \Delta^{(v,w)}$ with $\Delta^{(v,w)}$ given by

$$\Delta^{(v,w)} = \left(\Delta E_{\exp}^{(v,0)} - \Delta E_{\text{calc}}^{(v,0)}\right) + \left(\text{IP}_{\text{calc}}^{(w,0)} - \text{IP}_{\exp}^{(w,0)}\right) \quad (5.16)$$

Here, $\Delta E^{(v,0)}$ and $\text{IP}^{(w,0)}$ are the vertical excitation energies and ionization potentials at the Franck-Condon geometry, and the subscripts refer to experimental and calculated values. This correction ensures that the predicted kinetic energy of the photoelectrons ejected close to time zero from the initially excited state v by ionization to the state w matches the experimental value.

5.2 Non-Perturbative Wavepacket-Based Calculation

The approximate perturbative method of calculating the time-resolved photoelectron spectrum outlined in the previous section lends itself nicely to be used in conjunction with trajectory-based dynamics simulations. To go beyond the perturbative treatment, the field-matter interaction due to the pump and probe pulses has to be explicitly added to the molecular Hamiltonian, and the system propagated under the influence of this total Hamiltonian [7, 18]. We represent the pump and probe fields by

$$\mathcal{E}_j(t) = \frac{1}{2}\varepsilon_{j0}\exp[-4\ln(2)(t - t_j)^2/\tau_j^2]\left(\exp[i\omega_j t] + \exp[-i\omega_j t]\right) \qquad (5.17)$$

In the dipole approximation, the total Hamiltonian is given by

$$\hat{H}_{\text{tot}} = \hat{H} - \sum_j^{\text{pu,pr}} \hat{\mu} \cdot \mathcal{E}_j(t) \qquad (5.18)$$

In contrast to the previous method, the excitation and ionization processes are not assumed impulsive, and it is not assumed that the pulses are non-overlapping. The Hamiltonian is represented in the basis of the neutral electronic states as well as the different ionization continua corresponding to different cationic states. The state of the system is given according to

$$|\Psi(t)\rangle = \sum_v |\Phi^{(v)}(t)\rangle|\psi^{(v)}\rangle + \sum_w \int_0^\infty |\Phi^{(w)}(t)\rangle|\psi^{(w)}(E_k)\rangle \, \mathrm{d}E_k \qquad (5.19)$$

where v labels the neutral states, and w labels the cationic states or equivalently the ionization continua. The cationic electronic state $|\psi^{(w)}(E_k)\rangle$ is a function of the photoelectron kinetic energy E_k. As previously, the photoelectron spectrum can be calculated from the probability density of the ionization continua long after all fields have decayed, i.e. from the overlap between the state $|\Psi(t)\rangle$ for $t \to \infty$ and the electronic ionization continua

$$\sigma(E_k, \Delta t) = \sum_w |\langle\Psi(t \to \infty)|\psi^{(w)}(E_k)\rangle|^2 \qquad (5.20)$$

where $\Delta t = t_{\text{pr}} - t_{\text{pu}}$ is the delay between the center of the field envelopes of the pump and probe pulses. It should be clear that this approach in principle involves a number of simulations—one for each value of Δt—in contrast to the previously described method. Notice that unlike Eq. (5.2), the state of the photoelectron is not explicitly considered in the calculation.

To calculate the spectrum as given in Eq. (5.20), the state $|\Psi(t)\rangle$ is represented in the framework of Multi-Configuration Time-Dependent Hartree (MCTDH) and propagated following the standard prescription presented briefly in Sect. 4.1. The ionization continua are discretized such that the integral over kinetic energy is replaced by a sum. In essence, we include in the MCTDH calculations a number of discrete neutral states, and for each ionization continuum, we include a large number of identical cationic states displaced in energy to represent the variable energy of the photoelectron, Fig. 5.1. Thus, we make the approximation that the electron is ejected without interacting with the core of the cation such that the energies of the photoelectron and the cationic states are additive as also assumed in the previous method.

For the excitation and ionization processes, we invoke the Condon approximation, i.e. the dependence of the transition dipole moment on the nuclear coordinates is neglected. This approximation is somewhat justified as the MCTDH calculations

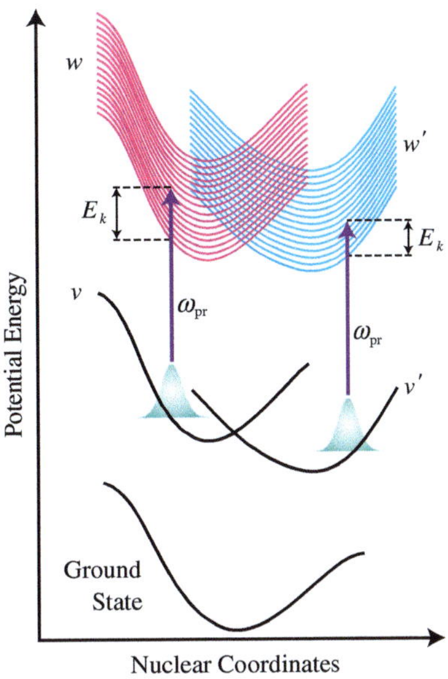

Fig. 5.1 Ionization of a nuclear wavepacket from two different neutral excited states v and v' to two different cationic states w and w' respectively. The ionization continua are represented by a set of discrete states displaced in energy to represent the energy of the ejected photoelectron E_k

are performed in the diabatic representation in which the electronic character of the states is not or only weakly dependent on the nuclear coordinates. In addition, the initial state is taken as the narrow vibrational ground state of the ground electronic state over the span of which the change in the transition dipole moment is negligible. Hereby, we neglect among others the Herzberg-Teller effect [19]. In contrast to the previously described method, the transition dipole moment involved in the photoionization process is also assumed to be independent of nuclear coordinates. Moreover, the transition dipole moment is assumed to be independent of the photoelectron kinetic energy over the small energy ranges studied inhere similar to the approach of Seel and Domcke [7]. This is in accordance with experimental results on ionization cross-sections [20, pp. 41–43]. In the simulations, this is achieved by representing the discretized continua by grids evenly spaced in energy.

References

1. R. Bersohn, A.H. Zewail, Ber. Bunsenges. Phys. Chem. **92**, 373–378 (1988)
2. J. Petersen, N.E. Henriksen, K.B. Møller, Chem. Phys. Lett. **539–540**, 234–238 (2012)
3. C. Meier, V. Engel, Phys. Rev. Lett. **73**, 3207–3210 (1994)
4. C. Meier, V. Engel, J. Chem. Phys. **101**, 2673–2677 (1994)
5. K.B. Møller, N.E. Henriksen, A.H. Zewail, J. Chem. Phys. **113**, 10477–10485 (2000)
6. S. Gräfe, D. Scheidel, V. Engel, N.E. Henriksen, K.B. Møller, J. Phys. Chem. A **108**, 8954–8960 (2004)
7. M. Seel, W. Domcke, J. Chem. Phys. **95**, 7806–7822 (1991)
8. G.A. Worth, R.E. Carley, H.H. Fielding, Chem. Phys. **338**, 220–227 (2007)

9. B.T. Pickup, O. Goscinski, Mol. Phys. **26**, 1013–1035 (1973)
10. B.T. Pickup, Chem. Phys. **19**, 193–208 (1977)
11. S. Patchkovskii, Z. Zhao, T. Brabec, D.M. Villeneuve, Phys. Rev. Lett. **97**, 123003 (2006)
12. Y. Öhrn, G. Born, Adv. Quantum Chem. **13**, 1–88 (1981)
13. I.G. Kaplan, B. Barbiellini, A. Bansil, Phys. Rev. B **68**, 235104 (2003)
14. C.M. Oana, A.I. Krylov, J. Chem. Phys. **127**, 234106 (2007)
15. C.M. Oana, A.I. Krylov, J. Chem. Phys. **131**, 124114 (2009)
16. L. Tao, C. M. Oana, V. A. Mozhayskiy, A. I. Krylov, ezDyson, http://iopenshell.usc.edu/downloads/ezdyson
17. J.R. Klauder, B.-S. Skagerstam, *Coherent States—Applications in Physics and Mathematical Physics* (World Scientific, Singapore, 1985)
18. M. Seel, W. Domcke, Chem. Phys. **151**, 59–72 (1991)
19. G. Herzberg, E. Teller, Z. Phys, Chem. B **21**, 410–446 (1933)
20. J.W. Rabalais, *Principles of Ultraviolet Photoelectron Spectroscopy* (John Wiley & Sons, New York, NY, 1977)

Chapter 6
Electronic Structure

The methods for simulating nuclear dynamics presented in the previous chapters rely on a very fundamental constituent—the potential energy surface. Potential energy surfaces can be obtained from spectroscopic measurements on the basis of which analytic potential energy functions can be fitted. Infrared and Raman spectroscopy can be used to obtain the electronic ground state potential energy surface. Laser-induced fluorescence and absorption spectroscopy can be used to determine the potential energy surfaces of the excited states if the ground state is well characterized [1, 2]. However, for all but the smallest molecules, the determination of full-dimensional potential energy surfaces from spectroscopic measurements is a monumental task. On the other hand, methods of electronic structure calculation can readily be applied to this challenge.

6.1 Coupled-Cluster Methods

To be applicable for our interests, a given electronic structure method should be able to describe the ground state as well as the excited states within the same framework. Furthermore, it is desirable if the method provides a hierarchy of approximations where the accuracy can be consistently improved upon by moving up in the hierarchy. One such approach is the coupled-cluster method [3, 4] in which properties such as excitation energies can be calculated using response theory [5–8]. In coupled-cluster theory, the ansatz for the electronic state is given by [9, pp. 650–654].

$$|\psi_{CC}\rangle = \exp(\hat{T})|\psi_0\rangle \tag{6.1}$$

where $|\psi_0\rangle$ is the reference usually taken as the Hartree-Fock state. $\hat{T}$ is the so-called cluster operator, not to be confused with the kinetic energy operator, given by

$$\hat{T} = \sum_j t_j \hat{\tau}_j \tag{6.2}$$

T. S. Kuhlman, *The Non-Ergodic Nature of Internal Conversion*, Springer Theses, DOI: 10.1007/978-3-319-00386-3_6, © Springer International Publishing Switzerland 2013

The general excitation operator $\hat{\tau}_j$ exchanges a given number of occupied spin-orbitals in the reference state with unoccupied spin-orbitals, or, equivalently, excites a given number of electrons from the reference state. The amplitude of each excitation is denoted by t_j. Truncation of the cluster operator at a given excitation level leads to a hierarchy of methods: CCS, CC2, CCSD, CC3 etc [10–12]. Furthermore, CCSDR(3), a non-iterative analog of CC3, can be defined which lies in-between CCSD and CC3 in accuracy [13]. In addition to response theory, the equation of motion (EOM) formalism can also be used to obtain excitation energies [14–16]. In the EOM formalism, ionized states can also be treated by use of EOMIP allowing for calculation of cationic potential energy surfaces in the same formalism as the neutral excited states [17]. The coupled-cluster calculations presented in Chap. 7 were performed using CFOUR [18] and DALTON [19].

6.2 Multi-Reference Methods

In cases of strong interaction between electronic states, it is necessary to use a method which is capable of describing regions of near degeneracy between potential energy surfaces in the vicinity of conical intersections. One such method is Multi-Configuration Self-Consistent Field (MCSCF) which corrects the reference state for static correlation effects. In MCSCF, the ansatz for the electronic state is given by

$$|\psi_{\mathrm{MC}}\rangle = \left(1 + \sum_j t_j \hat{\tau}_j\right)|\psi_0\rangle \tag{6.3}$$

with the same definitions for $\hat{\tau}_j$ and t_j as in coupled-cluster. In MCSCF, both the amplitudes t_j as well as the molecular orbitals making up the reference state $|\psi_0\rangle$ are optimized simultaneously [20–22]. To select which configurations to include in the summation in Eq. (6.3), the Complete-Active Space Self-Consistent Field (CASSCF) method is often invoked. In CASSCF, an active space consisting of both occupied and unoccupied orbitals is chosen, and within this subspace all configurations are included, i.e. full configuration interaction. A CASSCF calculation with n electrons distributed in m active orbitals is termed CAS(n, m)SCF. An important development for the simultaneous treatment of several states using CASSCF is state averaging. Instead of optimizing the orbitals for each state individually, the optimization is done in an averaged fashion for a specific number of states to ensure that they are treated on an equal footing. A state-averaged CASSCF calculation in which the orbitals are optimized simultaneously for N_e states is termed SA-N_e-CAS(n, m)SCF. The SA-CASSCF calculations presented in Chap. 9 were performed using MOLPRO 2010.1 [23].

CASSCF retrieves part of the static correlation. The so-called dynamical correlation can partly be retrieved by multi-reference perturbation theory using the CASSCF wavefunction as the reference usually in the form of CASPT2, i.e.

second-order perturbation theory [24–26]. CASPT2 can also be performed for multiple states yielding Multi-State Multi-Reference CASPT2 (MS-MR-CASPT2) when all states are treated together [27, 28]. One problem encountered in CASPT2 calculations is so-called intruder states, i.e. states which are not included in the CASSCF calculation but show up due to the subsequent perturbation theory treatment. This problem can be remedied by use of a level-shift [29]. The on-the-fly MS-MR-CASPT2 dynamics and time-resolved photoelectron spectra calculations presented in Chap. 8 were performed using the combined Ab Initio Multiple Spawning and MOLPRO 2006.2 code [30, 31]. These simulations also included calculation of MS-MR-CASPT2 analytic non-adiabatic couplings [32, 33]. For the calculation of time-resolved photoelectron spectra, cationic states were also treated, however, the orbitals from the calculation of the neutral states were frozen resulting in a CAS configuration interaction treatment of these states.

References

1. J. Laane, Int. Rev. Phys. Chem. **18**, 301–341 (1999)
2. J. Laane, J. Phys. Chem. A **104**, 7715–7733 (2000)
3. J. Čížek, J. Chem. Phys. **45**, 4256–4266 (1966)
4. R.J. Bartlett, Annu. Rev. Phys. Chem. **32**, 359–401 (1981)
5. H.J. Monkhorst, Int. J. Quantum Chem. **12**, 421–432 (1977)
6. D. Mukherjee, P.K. Mukherjee, Chem. Phys. **39**, 325–335 (1979)
7. H. Koch, P. Jørgensen, J. Chem. Phys. **93**, 3333–3344 (1990)
8. H. Koch, H.J.A. Jensen, P. Jørgensen, T. Helgaker, J. Chem. Phys. **93**, 3345–3350 (1990)
9. T. Helgaker, P. Jørgensen, J. Olsen, *Molecular Electronic-Structure Theory* (Wiley, Chichester, 2000)
10. O. Christiansen, H. Koch, P. Jørgensen, Chem. Phys. Lett. **243**, 409–418 (1995)
11. O. Christiansen, H. Koch, P. Jørgensen, J. Chem. Phys. **103**, 7429–7441 (1995)
12. H. Koch, O. Christiansen, P. Jørgensen, A.M.S. de Merás, T. Helgaker, J. Chem. Phys. **106**, 1808–1818 (1997)
13. O. Christiansen, H. Koch, P. Jørgensen, J. Chem. Phys. **105**, 1451–1459 (1996)
14. H. Sekino, R.J. Bartlett, Int. J. Quantum Chem. **26**, 255–265 (1984)
15. J.F. Stanton, R.J. Bartlett, J. Chem. Phys. **98**, 7029–7039 (1993)
16. D.C. Comeau, R.J. Bartlett, Chem. Phys. Lett. **207**, 414–423 (1993)
17. J.F. Stanton, J. Gauss, J. Chem. Phys. **101**, 8938–8944 (1994)
18. Cfour, a quantum chemical program package written by J.F. Stanton et al. For the current version http://www.cfour.de
19. Dalton, a molecular electronic structure program, Release2.0 (2005) http://daltonprogram.org
20. H.-J. Werner, P.J. Knowles, J. Chem. Phys. **82**, 5053–5063 (1985)
21. P.J. Knowles, H.-J. Werner, Chem. Phys. Lett. **115**, 259–267 (1985)
22. M.W. Schmidt, M.S. Gordon, Annu. Rev. Phys. Chem. **49**, 233–266 (1998)
23. H.-J. Werner, P.J. Knowles, R. Lindh, F.R. Manby, M. Schütz, Molpro, version 2010.1, a package of ab initio programs and others http://www.molpro.net
24. K. Andersson, P.-Å. Malmqvist, B.O. Roos, J. Chem. Phys. **96**, 1218–1226 (1992)
25. B.O. Roos, K. Andersson, M.P. Fülscher, P.-Å. Malmqvist, L. Serrano-Andrés, P. Pierloot, M. Merchán, Adv. Chem. Phys. **93**, 219–331 (1996)
26. H.-J. Werner, Mol. Phys. **89**, 645–661 (1996)
27. J. Finley, P.-Å. Malmqvist, B.O. Roos, L. Serrano-Andrés, Chem. Phys. Lett. **288**, 299–306 (1998)

28. T. Shiozaki, W. Győrffy, P. Celani, H.-J. Werner, J. Chem. Phys. **135**, 081106 (2011)
29. B.O. Roos, K. Andersson, Chem. Phys. Lett. **245**, 215–223 (1995)
30. B.G. Levine, J.D. Coe, A.M. Virshup, T.J. Martínez, Chem. Phys. **347**, 3–16 (2008)
31. H.-J. Werner, P.J. Knowles, R. Lindh, F.R. Manby, M. Schütz, Molpro version 2006.2, a package of ab initioprograms and others http://www.molpro.net
32. T. Mori, S. Kato, Chem. Phys. Lett. **476**, 97–100 (2009)
33. T. Mori, W.J. Glover, M.S. Schuurman, T.J. Martínez, J. Phys. Chem. A **116**, 2808–2818 (2012)

Part IV
Results and Discussion

Chapter 7
The Cycloketones

The carbonyl chromophore plays a central role in organic photochemistry and photophysics in particular due to the interplay between two types of excited states resulting from the promotion of an electron to the anti-bonding π^* orbital from either the π or an n orbital [1]. Prominent examples of processes involving these states is the Norrish Type I α-cleavage [2–5], the Norrish Type II intramolecular γ-hydrogen abstraction, and efficient intersystem crossing with a simultaneous change in the orbital angular momentum as established by El-Sayed's selection rule [6–8].

One group of molecules incorporating the carbonyl chromophore is the cycloketones. In addition to the involvement in the processes described above, we will show inhere that the cycloketones also represent a very good model system to exhibit the subtle details of an internal conversion process. By use of time-resolved mass spectrometry (TRMS) and photoelectron spectroscopy (TRPES) as well as wavepacket simulations, we investigate the dynamical nature of the $S_2 \rightarrow S_1$ internal conversion—at the Franck-Condon geometry corresponding to the $(n, 3s) \rightarrow (n, \pi^*)$ transition. This transition from a molecular Rydberg state to a valence excited state involves two states of significantly different electronic character and thereby represents an example of weak interaction.

The study involves the seven cycloketones presented in Fig. 7.1. The molecules cover three ring sizes from four to six carbon atoms as well as methyl and ethyl substitutions in different positions on the ring. Whereas cyclobutanone and cyclohexanone belong to the C_s point group and cyclopentanone belongs to the C_{2v} point group, the substituted molecules do not posses any symmetry. These variations between the molecules allow us to investigate the influence of a range of structurally, and to some extent electronically, determined parameters on the rate of internal conversion: the C-CO-C angle and ring strain, the specific nature and frequency of vibrational motion, the vibrational density of states, and point group symmetry. In contrast to the variations in structure, the energy of the (n, π^*) and $(n, 3s)$ states at the Franck-Condon geometry do not differ significantly between the seven molecules in particular not for molecules of the same ring size. The absorption maximum is found at

T. S. Kuhlman, *The Non-Ergodic Nature of Internal Conversion*, Springer Theses, DOI: 10.1007/978-3-319-00386-3_7, © Springer International Publishing Switzerland 2013

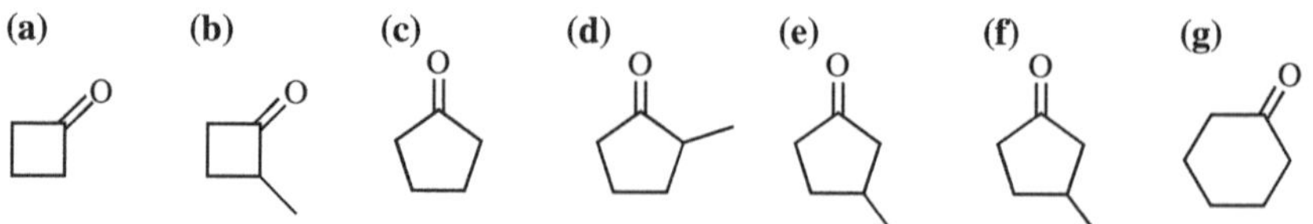

Fig. 7.1 The seven cycloketones included in the study. (**a**) cyclobutanone (CB), (**b**) 2-methyl-cyclobutanone (2-MeCB), (**c**) cyclopentanone (CP), (**d**) 2-methylcyclopentanone (2-MeCP), (**e**) 3-methylcyclopentanone (3-MeCP), (**f**) 3-ethylcyclopentanone (3-EtCP), and (**g**) cyclohexanone (CH)

Table 7.1 Details of the pump and probe pulses used in the time-resolved experiments on the cycloketones

Molecules	ω_{pu}/eV	E_{pu}/nJ	ω_{pr}/eV	E_{pr}/mJ	τ_{xc}/fs
CB	6.20	350	3.52	2.0	124
2-MeCB	6.20	230	3.73	1.9	172
CP	6.20	350	3.52	2.0	124
2-MeCP	6.20	230	3.73	1.9	179
3-MeCP	6.20	230	3.73	1.9	172
3-EtCP	6.20	230	3.73	1.9	172
CH	6.33	120	3.52	2.0	126

4.28 ± 0.15 eV for the (n, π^*) state and at 6.20 ± 0.10 eV for the $(n, 3s)$ state [9–13]. This similarity allows for a direct comparison between the molecules.

7.1 Time-Resolved Experiments

Parts of the results described in this section have been published in [14, 15]. Some text passages and/or figures and tables in this section have been reproduced and adapted with permission from John Wiley and Sons.

In the time-resolved experiments, the molecules were excited directly by one pump photon to the $(n, 3s)$ state from which they can be ionized by a single probe photon. Upon internal conversion to the (n, π^*) state, the energy of one probe photon is not sufficient to ionize the molecule upon conservation of nuclear kinetic energy. Table 7.1 lists the typical frequencies and energies of the pump and probe pulses as well as their cross-correlation (XC) determined experimentally.

The temporal evolution of the parent ion currents following excitation to the $(n, 3s)$ state is presented in Figs. 7.2 and 7.3. In four cases, the time-resolved photoelectron spectra were also recorded giving rise to the transients of the integrated $(n, 3s)$ photoelectron band presented in Fig. 7.4. The temporal evolution of the parent ion currents closely resemble those of the $(n, 3s)$ photoelectron band for these four cases. Consequently, the decay in the ion yield can be used as a measure of the

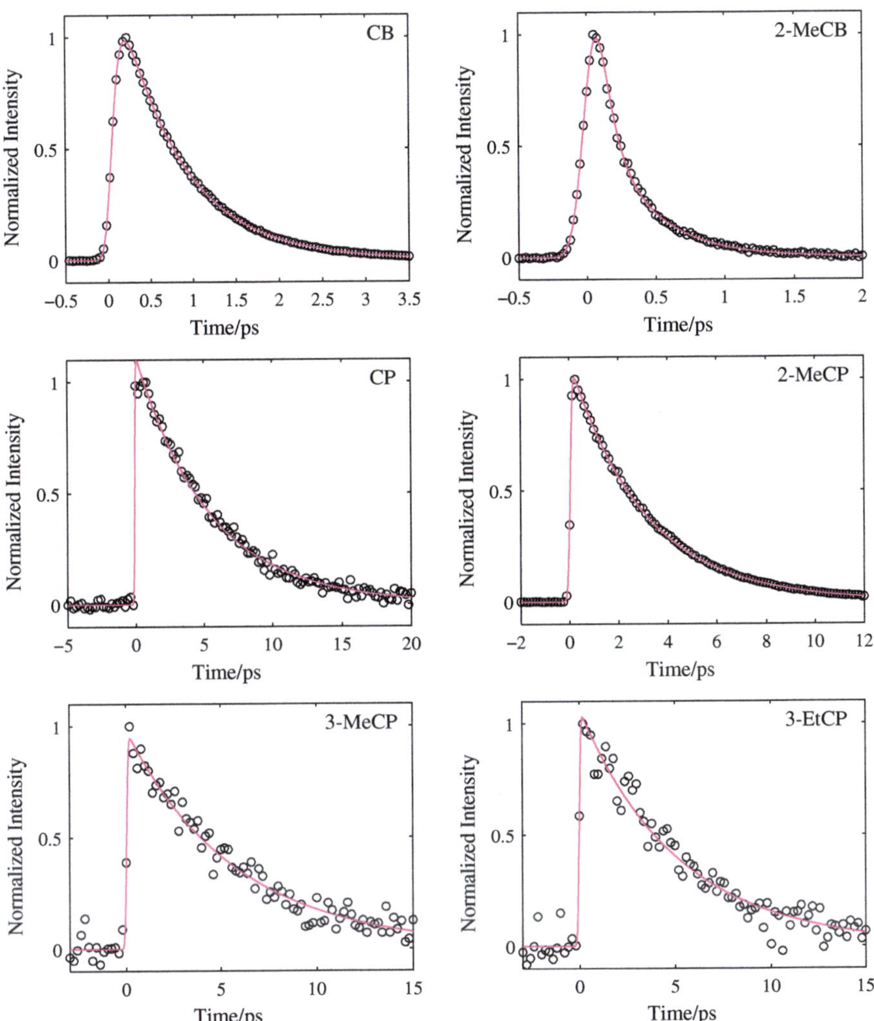

Fig. 7.2 Temporal evolution of the parent ion currents for six of the seven cycloketones following excitation to the $(n, 3s)$ state. For cyclopentanone, 3-methylcyclopentanone, and 3-ethylcyclopentanone, a non-resonant XC component has been subtracted from the signal and fit. Adapted with permission from Kuhlman et al. [14, 15]. Copyright John Wiley and Sons

lifetime of the $(n, 3s)$ state. This correspondence between ion and photoelectron signals is a consequence of the chosen pump-probe scheme. The ion currents reveal a set of timescales for the $(n, 3s) \rightarrow (n, \pi^*)$ transition, i.e. $S_2 \rightarrow S_1$, ranging over more than an order of magnitude from 0.37 ± 0.01 ps for 2-methylcyclobutanone to 9.67 ± 0.43 ps for cyclohexanone, cf. Table 7.2. A similar trend is observed for the $(n, 3s)$ photoelectron decays, cf. Table 7.3. Two clear trends are observed in these timescales: (1) the timescale increases with increasing ring size, and (2) substitution

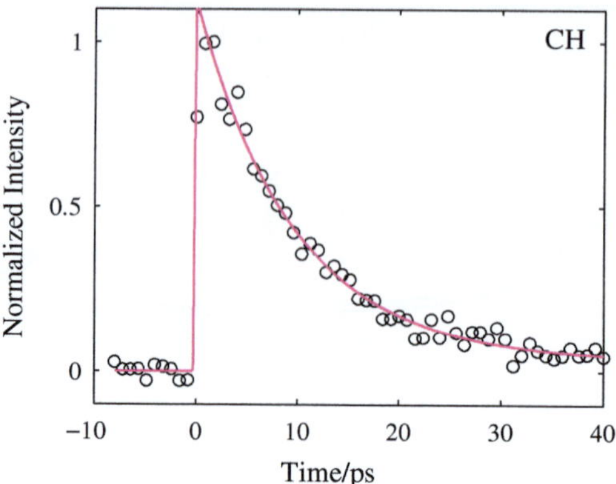

Fig. 7.3 Temporal evolution of the parent ion current for cyclohexanone following excitation to the $(n, 3s)$ state. A non-resonant XC component has been subtracted from the signal and fit

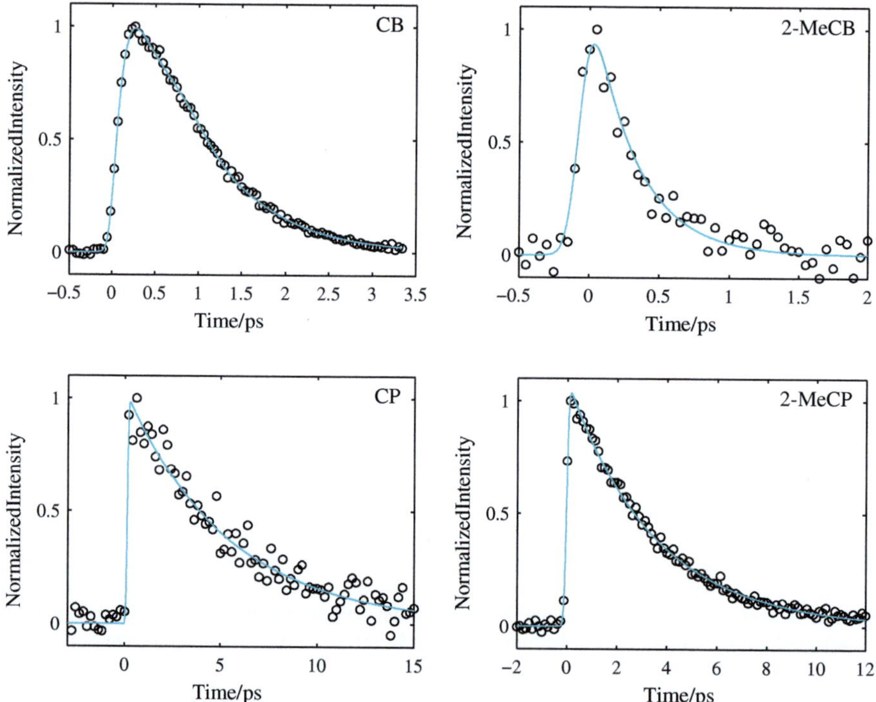

Fig. 7.4 Temporal evolution of the photoelectron currents of the spectrally integrated $(n, 3s)$ photoelectron band. For cyclopentanone, a non-resonant XC component has been subtracted from the signal and fit. Adapted with permission from Kuhlman et al. [14, 15]. Copyright John Wiley and Sons

Table 7.2 Timescales for the decay of the parent ion current for the seven cycloketones following excitation to the $(n, 3s)$ state

Molecule	τ_1/ps	τ_2/ps
CB	0.08 ± 0.01	0.74 ± 0.01
2-MeCB	0.08 ± 0.01	0.37 ± 0.01
CP		5.39 ± 0.17
2-MeCP		3.05 ± 0.01
3-MeCP		5.79 ± 0.16
3-EtCP		5.16 ± 0.17
CH		9.67 ± 0.43

Reprinted with permission from Kuhlman et al. [14, 15]. Copyright John Wiley and Sons

Table 7.3 Timescales for the decay of the $(n, 3s)$ photoelectron band for four of the cycloketones. Also given in two cases is the period of the spectral oscillation of the band

Molecule	τ_1/ps	τ_2/ps	T/ps
CB	0.31 ± 0.06	0.74 ± 0.02	0.95 ± 0.03^a
2-MeCB		0.32 ± 0.02	
CP		5.37 ± 0.11	
2-MeCP		3.47 ± 0.03	0.42 ± 0.01^b

Reprinted with permission from Kuhlman et al. [15]. Copyright John Wiley and Sons
[a] From the fit to the low-energy part of the $(n, 3s)$ photoelectron band
[b] Average period of oscillation from the fit to the low- and high-energy part of the $(n, 3s)$ photoelectron band

in the 2-position but not in the 3-position decreases the timescale. According to the standard energy gap law derived from Fermi's golden rule for the non-radiative transition rate, the rate of transition should be an exponential decaying function of the energy gap between the two states involved [16, 17]. As the energy gap between the $(n, 3s)$ and (n, π^*) states does not differ significantly between the molecules, such a law does not straightforwardly explain the large timescale differences observed here. We will inhere disentangle the dynamics of the internal conversion process in the cycloketones and thereby identify the specific properties which can explain the inter- and intragroup timescale differences observed.

TRPES provides more information than the timescale for electronic population transfer. Figure 7.5 depicts the transients of the spectrally integrated $(n, 3s)$ photoelectron band of cyclobutanone and 2-methylcyclopentanone. The integration has been performed over the low- and high-energy spectral halves of the band separately. Such an integration scheme reveals spectrally oscillating features—an unequivocal sign of coherent nuclear motion affecting the electronic structure. For cyclobutanone, the period of oscillation is 0.94 ± 0.03 ps whereas it is 0.42 ± 0.01 ps in the case of 2-methylcyclopentanone. These periods correspond to frequencies of $\sim$4 meV ($\sim$35 cm^{-1}) and $\sim$10 meV ($\sim$80 cm^{-1}) respectively. Due to the very low frequency, in particular in the case of cyclobutanone, the observations reveal exactly which vibra-

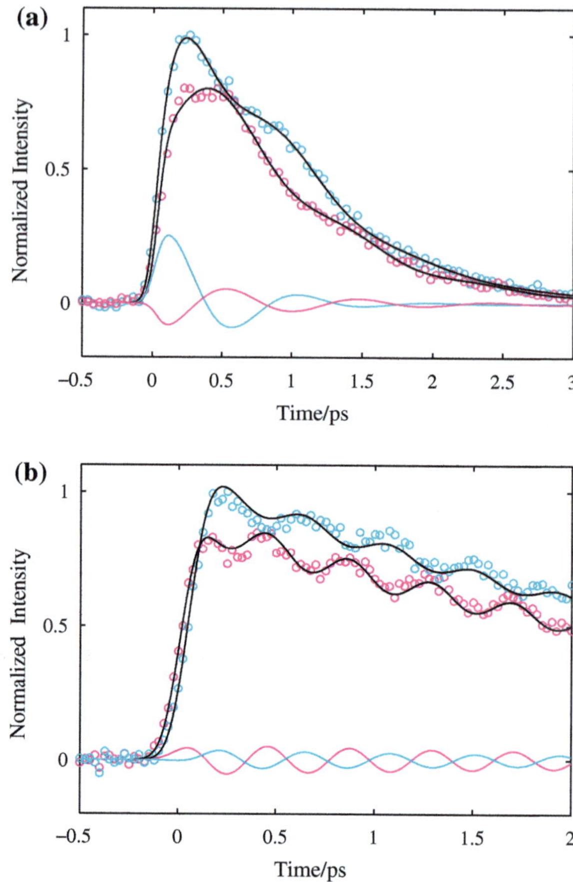

Fig. 7.5 Transients of the low- (blue) and high-energy (pink) spectral halves of the $(n, 3s)$ photoelectron band along with fits ($-$). In both cases, a clear phase relationship is observed between the oscillatory components of the fits indicated by the respective *colored lines*. (**a**) Cyclobutanone. (**b**) 3-methylcyclopentanone—reprinted with permission from Kuhlman et al. [15]. Copyright John Wiley and Sons

tional mode is dominant in mediating the coupling between the two electronic states. In the ground state, this low-frequency ring-puckering mode primarily involves the C-CO-C moiety of the molecule with the carbonyl group bending out of the molecular plane albeit with some motion of the last CH_2 group in the case of cyclobutanone [18–20]. The large difference in frequency between the two molecules is an important factor in explaining the observed differing timescales, however, it is not fully sufficient on its own. As we will show below, at least two factors have to be taken into account: the frequency of the specific vibrational mode involved and the difference in energy of the $(n, 3s)$ state between the Franck-Condon and equilibrium geometries, i.e. the difference between the vertical and the adiabatic excitation energy. In addition, the total density of vibrational states on the (n, π^*) surface plays a small role.

7.1.1 A Representative Model

The differences in the timescale of the $(n, 3s) \rightarrow (n, \pi^*)$ transition in the seven cycloketones can be rationalized by considering the model depicted in Fig. 7.6a. Once excited to the $(n, 3s)$ state, the molecule will vibrate in some modes, and based on the TRPES data, it can be concluded that only one (or a few) of these plays an important role in the pathway to the lower electronic state. We will therefore restrict the discussion to a one-dimensional representation. In one dimension, the position of the crossing point with the lower electronic surface (or possibly avoided crossing) is given by two factors (everything else being equal): the frequency of the vibrational mode in question and the energy difference between the Franck-Condon and equilibrium geometries of the $(n, 3s)$ state. As indicated in Fig. 7.6b, a low frequency, i.e. a small curvature, and a large energy difference will allow the molecule to access a larger configurational space whereby it can more easily access the region near the very important conical intersection facilitating a faster non-adiabatic population transfer. Such a transition is illustrated by the magenta arrow in Fig. 7.6a as opposed to the adiabatic dynamics indicated by the blue arrow.

The cause of the intergroup timescale differences, i.e. the differences in timescale between molecules of different ring size, is primarily rooted in the energy difference factor which is apparent by comparison of the unsubstituted cycloketones. The smaller strained cyclobutanone is able to relieve ring strain in the $(n, 3s)$ state through

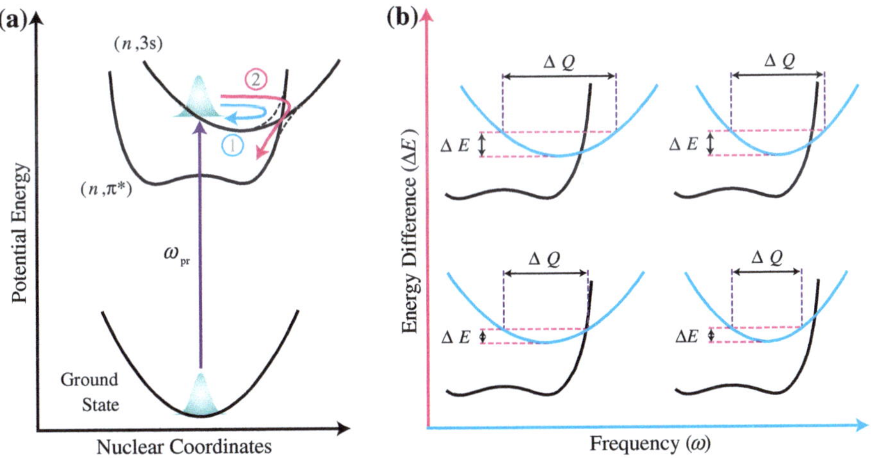

Fig. 7.6 (**a**) Scheme illustrating the possible dynamics following excitation to the $(n, 3s)$ state in the cycloketones. Depending on the configurational space sampled, the wavepacket can follow two pathways: ① an adiabatic pathway indicated by the *blue arrow* and ② a non-adiabatic pathway indicated by the *magenta arrow*. (**b**) The vibrational frequency ω and the energy difference ΔE between the Franck-Condon and equilibrium geometries of the $(n, 3s)$ state determine the intersection point between the two excited states and the configurational range available to the wavepacket ΔQ. Thereby, these two parameters significantly influence the rate of non-adiabatic transition

vibration in the ring-puckering mode whereas the five-membered ring of cyclopentanone is less prone to such motion. As calculated by EOM-CCSD, the energy difference in the $(n, 3s)$ state between the Franck-Condon and equilibrium geometries is 0.32 eV for cyclobutanone whereas it is only 0.14 eV for cyclopentanone. The more vibrationally congested $(n, 3s)$ absorption spectrum for cyclobutanone compared to cyclopentanone further corroborates this difference [12]. The even slower transition in cyclohexanone can be understood in terms of the inverse relationship between ring size and intensity of vibrational bands and, thus, release of angle strain in the C-CO-C moiety [10]. The frequency factor, caused by different curvatures of the potential energy surface as illustrated by the examples in Fig. 7.6b, also contributes to the intergroup timescale differences. The effect is clearly demonstrated by the observed anti-correlation between the vibrational frequency and the rate of transition for cyclobutanone and 2-methylcyclopentanone.

The intragroup timescale differences, i.e. the differences in timescale between molecules of the same ring size, can largely be understood in terms of the frequency factor and to what extent this is affected by substitution. Alkyl substitution in the 2-position leads to a significantly increased rate of transition whereas this is not the case for substitution in the 3-position. This is observed when comparing 2-methylcyclopentanone and 3-methylcyclopentanone. Also in cyclobutanone, substitution in the 2-position leads to an increased rate of transition. The vibrational mode central to the internal conversion process primarily involves motion in the C-CO-C moiety, thus, substitution in the 2-position should have a larger effect on the rate of transition compared to the 3-position as indeed is the case. This observation in turn cements the conclusion on the non-ergodicity of the process—i.e. the dynamics are truly localized not only in phase space, as deduced from the observation of coherent nuclear motion, but also in real space. The intergroup timescale differences also reflect this locality as the apparent non-local change of ring size actually has a very large local effect by significantly affecting the angle of the central C-CO-C moiety.

In identifying the pertinent mode mediating the population transfer, we have not discussed the role of point group symmetry. In cyclobutanone and cyclohexanone, the modes that couple the $(n, 3s)$ and (n, π^*) states are Franck-Condon active whereas this is not the case in cyclopentanone. By substitution in cyclopentanone, we break the C_{2v} symmetry of the molecule thereby lifting the symmetry restrictions on the coupling and Franck-Condon active modes. However, only in the case of substitution in the 2-position does this lead to a significant change in the rate of transition. This observation could speak against the importance of point group symmetry in these cases. The application of point group symmetry does not allow for an assessment of the size of a given matrix element, it only allows one to assess whether the matrix element is zero by symmetry. Thus, breaking of symmetry by substitution in the 2- and 3-position does not necessarily have the same consequence. Group theory only tells us that upon substitution in either position, certain matrix elements are not necessarily zero by symmetry anymore, however, they could still be negligible. Therefore, it is difficult to assess whether the increased rate of transition in cyclopentanone due to substitution in the 2-position is partly influenced by symmetry breaking in particular because a similar increase in transition rate is observed upon substitution in cyclobutanone which does not have the same symmetry restrictions. We will go into more detail regarding point group symmetry in Sect. 7.2.2.

Although it has been stressed that the dynamics leading to disposal of the electronic energy are truly localized, an increase in the total density of vibrational states on the lower surface does slightly speed up the process. This effect is a consequence of additional vibrational degrees of freedom acting as acceptor modes in the lower electronic state. Comparison between the rates of transition for the molecules of different ring size does not immediately reveal this aspect as the two other effects discussed play a much larger role. It is, however, revealed by a comparison between the rates of transition for 3-methylcyclopentanone and 3-ethylcyclopentanone. The addition of an extra CH_2-group increases the density of vibrational states by a factor of ~ 100 at an energy of $2\,eV$, approximately the energy difference between the $(n, 3s)$ and (n, π^*) states, as calculated using a Beyer-Swinehart algorithm [21]. However, this factor of 100 only leads to a small decrease in the timescale for transition from $5.79 \pm 0.16\,ps$ to $5.16 \pm 0.17\,ps$—very different than the behavior expected from application of theory in the statistical limit [16, 17, 22, 23].

7.2 Wavepacket Simulations

Some text passages and/or figures and tables in this section have been reprinted with permission from Kuhlman et al. [24]. Copyright 2012, American Institute of Physics.

The previous section provided a qualitative model for describing the $(n, 3s) \rightarrow (n, \pi^*)$ internal conversion in the cycloketones based on time-resolved experiments. In the next sections, results from wavepacket simulations on cyclobutanone and cyclopentanone will allow for a more in detail analysis of the process. By calculation of time-resolved photoelectron spectra, a basis for direct comparison between theory and experiment is provided. We will start out by briefly presenting the model Hamiltonian employed in the wavepacket calculations.

7.2.1 Model Hamiltonian

The ground and excited state equilibrium geometries of cyclobutanone and cyclopentanone are presented in Fig. 7.7. Structural parameters which differ significantly between the geometries are indicated. Upon population of the excited states, we expect nuclear motion towards the respective equilibrium geometry to take place. Thus, the model Hamiltonian employed in the wavepacket calculations should be given in terms of coordinates which can describe the changes between the ground and excited state equilibrium geometries. Model vibronic coupling Hamiltonians (VCHAM) were therefore parameterized in terms of five normal modes of vibration. In order of increasing frequency as calculated at the MP2/cc-pVTZ [25] level

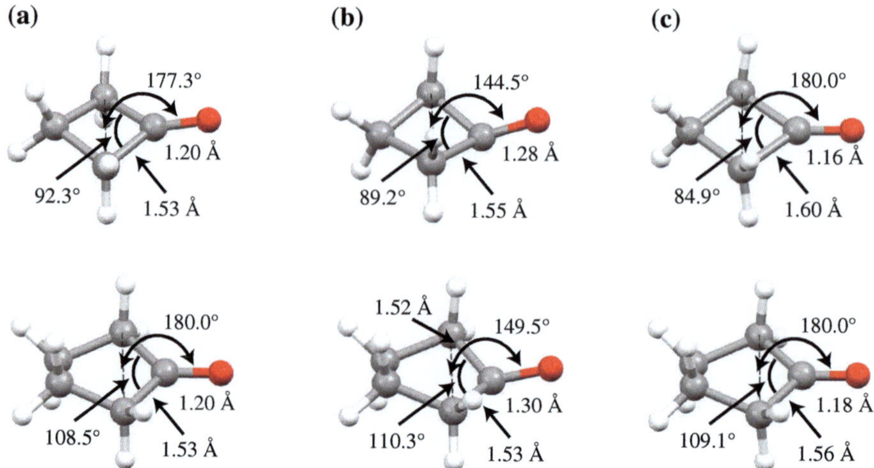

Fig. 7.7 Equilibrium geometries of the ground, (n, π^*), and $(n, 3\mathrm{s})$ states of cyclobutanone and cyclopentanone as obtained by CCSD/cc-pVTZ and EOM-CCSD/cc-pVTZ+1s1p1d. Important structural differences are indicated. Adapted with permission from Kuhlman et al. [24]. Copyright 2012, American Institute of Physics. (**a**) Ground state. (**b**) (n, π^*) state. (**c**) $(n, 3\mathrm{s})$ state

of theory, these modes are: ring-puckering, C=O out-of-plane bend (carbonyl pyramidalization), symmetric C-CO-C stretch, asymmetric C-CO-C stretch, and C=O stretch. The dimensionless normal coordinates Q_κ corresponding to these modes are labeled in order of increasing frequency as $\kappa = 1, 2, 7, 12, 21$ for cyclobutanone and $\kappa = 1, 3, 8, 16, 28$ for cyclopentanone.

The *ab initio* data used to fit the VCHAMs consisted of 1471 points for cyclobutanone and 1589 points for cyclopentanone calculated at the CCSD/EOM-CCSD level of theory with a cc-pVTZ+1s1p1d basis set. The 1s1p1d set of diffuse functions is described in Sect. A.1 of Appendix A. The ground, (n, π^*), $(n, 3\mathrm{s})$, and $(n, 3\mathrm{p})$ states as well as the cationic ground state were included in the calculations. The diabatic electronic states were numbered in order of increasing energy at the Franck-Condon point as $v = 1, 2, 3, 4, 5$ respectively. The two ground states were assumed uncoupled from the excited states except through the time-dependent fields of the pump and probe pulses. Examples of the fits obtained to the *ab initio* adiabatic potential energy surfaces are illustrated in Figs. 7.8 and 7.9. All parameters of the VCHAMs as well as the equilibrium geometries are tabulated in Appendix A.

7.2.2 Population Transfer Dynamics

We have performed wavepacket calculations using the VCHAMs for cyclobutanone and cyclopentanone in the Multi-Configuration Time-Dependent Hartree formalism (MCDTH). In one calculation, no time-dependent fields were included. The initial

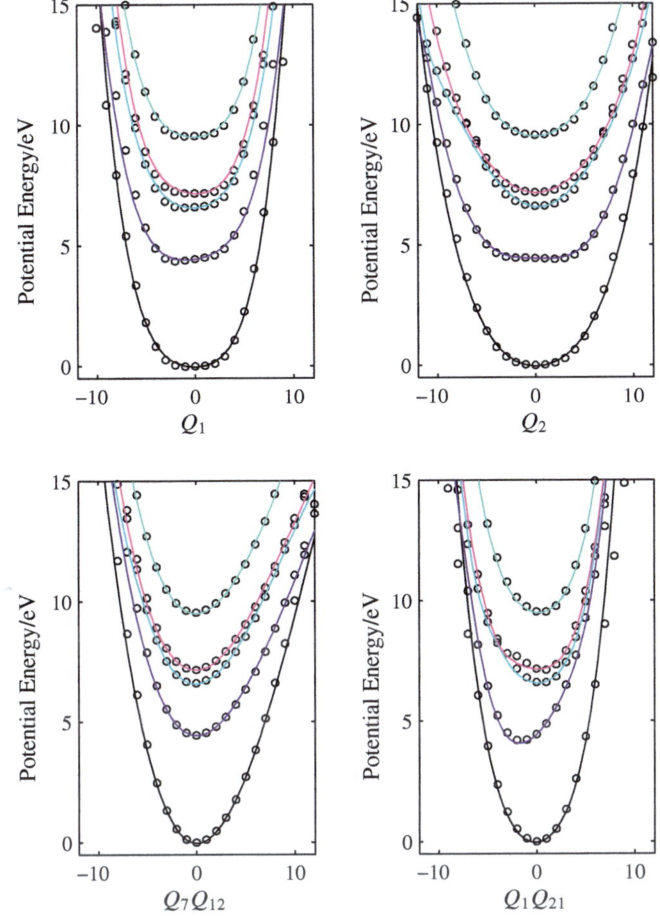

Fig. 7.8 Examples of CCSD/EOM-CCSD data (○) for cyclobutanone along with the VCHAM fits to the ground (black), (n, π^*) (purple), $(n, 3s)$ (blue), and $(n, 3p)$ (pink) states as well as the cationic ground state (green)

wavepacket in the $(n, 3s)$ state was taken as the Franck-Condon wavepacket obtained by operating with a unit dipole operator on the ground state wavefunction obtained by propagation in imaginary time [26]. In other calculations, either the pump field or both the pump and the probe fields of the form given in Eq. (5.17) were included. The Condon approximation was invoked in all cases, however, the electronic transition dipole moments calculated by linear response CCSD were used for the transitions to the excited states induced by the pump. For transitions from the excited states to the ground cationic state induced by the probe field, the electronic transition dipole moment was set to unity. The details of the basis sets, transition dipole moments, and fields are given in Tables 7.4, 7.5, and 7.6. To represent the ionization continuum, a

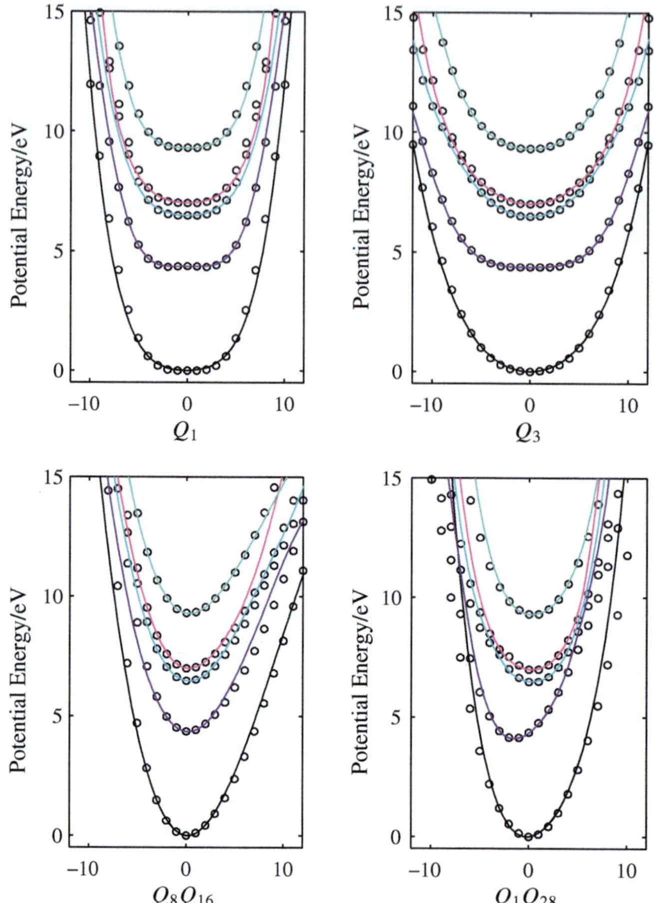

Fig. 7.9 Examples of CCSD/EOM-CCSD data (o) for cyclopentanone along with the VCHAM fits to the ground (black), (n, π^*) (purple), $(n, 3s)$ (blue), and $(n, 3p)$ (pink) states as well as the cationic ground state (green)

sine discrete variable representation (DVR) consisting of 151 equally spaced points from $E_k = 0.0$–1.5 eV was employed along with five single-particle functions. All calculations used a timestep of 0.2 fs and employed the variable mean-field integration scheme with a sixth-order Adams-Bashforth-Moulton predictor-corrector integrator and an error tolerance of 10^{-8}.

Figure 7.10 presents the absorption spectra obtained from both the calculations with and without fields. For both molecules, an intense peak is observed at a detuning from the vertical transition frequency of $\Delta\omega = -0.05$ eV. The spectrum is slightly more vibrationally congested in the case of cyclobutanone which is also observed experimentally albeit to a much larger extent [12].

Table 7.4 Parameters of the pump and probe fields used in the calculations on cyclobutanone/cyclopentanone $\Delta\omega$ indicates a possible detuning from the vertical transition frequency (as obtained from the VCHAM)

Field	ε_0/au	ω/eV	t_0/fs	τ/fs
Pump	$3.462 \cdot 10^{-3}$	$6.597/6.480 + \Delta\omega$	0	88
Probe	$7.555 \cdot 10^{-3}$	3.520	Δt	88

Table 7.5 Length gauge transition dipole moment, oscillator strength, and polarization of transitions from the ground state calculated by linear response CCSD/cc-pVTZ + 1s1p1d for cyclobutanone/cyclopentanone

State	μ/au	f	Pol.[a]
(n, π^*)	$6.2 \cdot 10^{-5}/8.7 \cdot 10^{-4}$	$6.8 \cdot 10^{-6}/9.4 \cdot 10^{-5}$	$(\mathbf{x}, \mathbf{z})/(\mathbf{z})$
$(n, 3\mathrm{s})$	$2.6 \cdot 10^{-1}/1.6 \cdot 10^{-1}$	$4.3 \cdot 10^{-2}/2.5 \cdot 10^{-2}$	$(\mathbf{x}, \mathbf{z})/(\mathbf{x}, \mathbf{y})$
$(n, 3\mathrm{p})$	$1.0 \cdot 10^{-2}/5.7 \cdot 10^{-4}$	$1.8 \cdot 10^{-3}/9.8 \cdot 10^{-4}$	$(\mathbf{x}, \mathbf{z})/(\mathbf{z})$

[a]$\mathbf{x}$: $\perp$ to molecular plane, $\mathbf{y}$: $\parallel$ to plane, and $\perp$ to C=O, and $\mathbf{z}$: $\parallel$ to plane, and $\parallel$ to C=O

Table 7.6 Number of single-particle functions for each mode and electronic state as well as the size of the harmonic oscillator DVR grid for cyclobutanone/cyclopentanone for the calculations without fields

κ	Ground	(n, π^*)	$(n, 3\mathrm{s})$	$(n, 3\mathrm{p})$	Cation	DVR grid
1/1	1 (3)	8	4	3	1 (3)	60/60
2/3	1 (3)	8	4	3	1 (3)	55/55
7/8	1 (3)	8	4	3	1 (3)	80/60
12/16	1 (3)	8	4	3	1 (3)	100/100
21/28	1 (3)	8	4	3	1 (3)	170/170

The corresponding values for the calculations including fields are given in parentheses if these differ. Adapted with permission from Kuhlman et al. [24]. Copyright 2012, American Institute of Physics

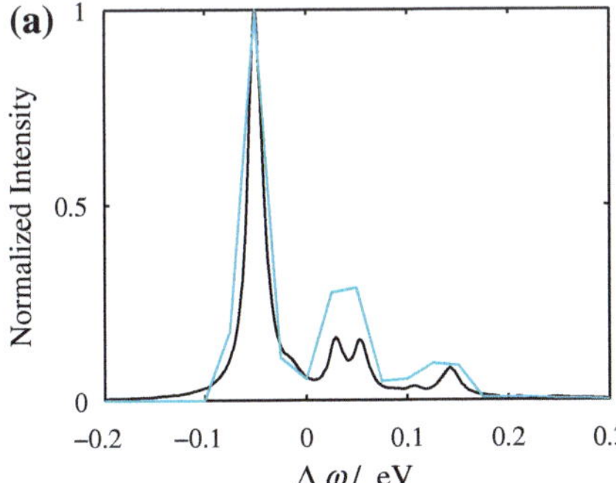
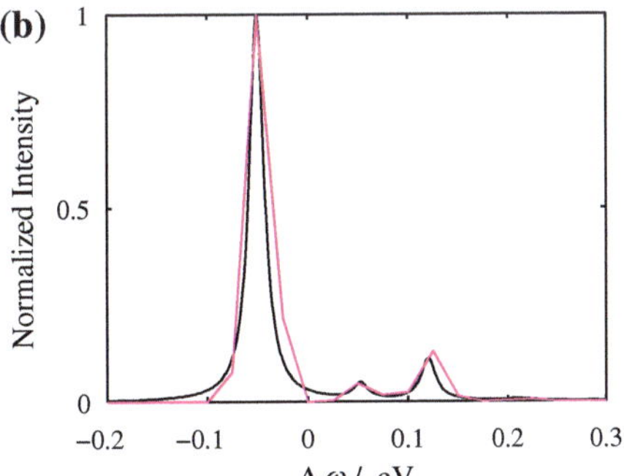

Fig. 7.10 Absorption spectra for the $(n, 3\mathrm{s})$ state. The intensity at a given detuning ($\Delta\omega$) from the vertical transition energy is given as the population of the $(n, 3\mathrm{s})$ state at $t = 100$ fs in a calculation with $\omega_{\mathrm{pu}} = \omega + \Delta\omega$. The *black lines* indicate spectra obtained by Fourier transformation of the autocorrelation function following a Franck-Condon excitation to the $(n, 3\mathrm{s})$ state (i.e. no fields included). (**a**) Cyclobutanone. (**b**) Cyclopentanone

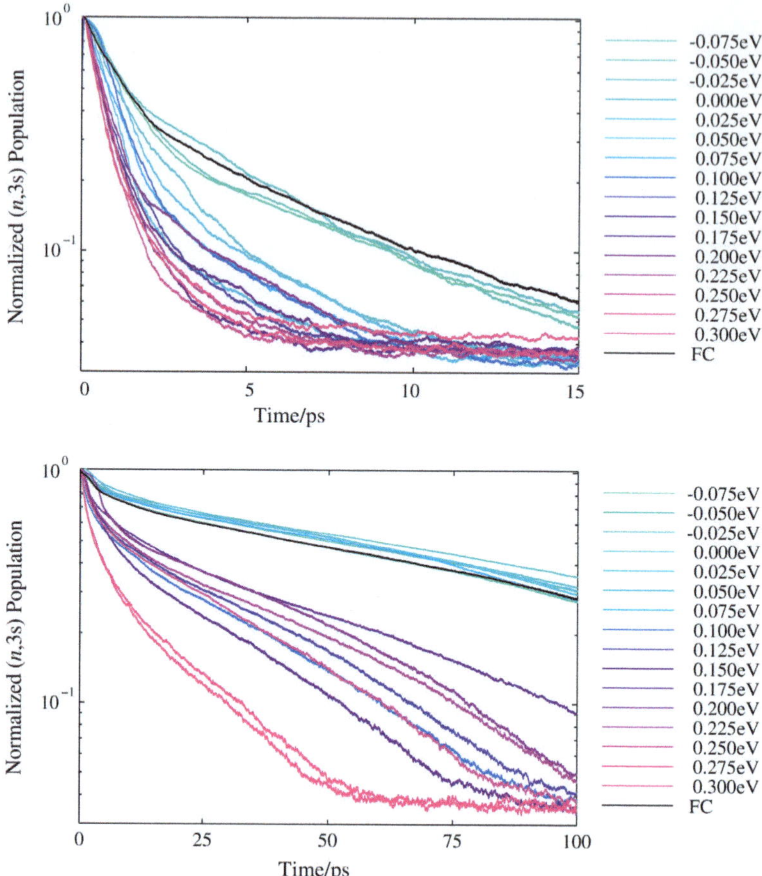

Fig. 7.11 Population of the $(n, 3s)$ state following excitation by a pump pulse with center frequency at different detunings $(\Delta\omega)$ from the vertical transition frequency as well as for the Franck-Condon (FC) excitation. The population has been normalized at $t = 100$ fs in all cases. (**a**) Cyclobutanone. (**b**) Cyclopentanone

The timescale for the $(n, 3s) \rightarrow (n, \pi^*)$ transition is dependent on the detuning as exhibited by the $(n, 3s)$ population decays in Fig. 7.11. A clear increase in the rate of transition is observed upon increasing the center frequency of the pump field from the intense peak at $\Delta\omega = -0.05\,\text{eV}$ to the higher energy vibrational peaks in the absorption spectra. For both molecules, we observe (at least) two components in the population decay: an initial prompt component with a fast decay rate and a delayed component with a slower decay rate. Higher excitation energy shifts the ratio of the two components towards the prompt decay component. The ratio of the prompt to the delayed component is much larger for cyclobutanone compared to cyclopentanone which, in addition to slower decay rates for the latter, leads to the longer overall timescale for population decay in cyclopentanone compared to

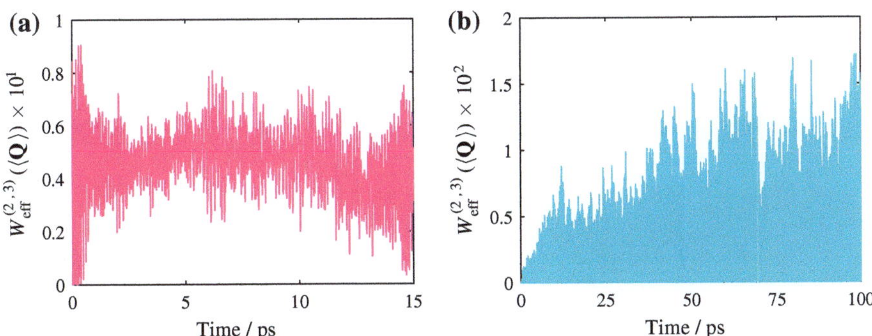

Fig. 7.12 Effective diabatic coupling between the (n, π^*) and $(n, 3s)$ states. Notice the different scaling of the ordinate axis for (**a**) cyclobutanone and (**b**) cyclopentanone

cyclobutanone. It should be clear from Fig. 7.11b that the timescale for population transfer in the simulations on cyclopentanone is much longer than what is observed in the experiments.

The large difference in the timescale for population transfer between cyclobutanone and cyclopentanone can be rationalized on the basis of the effective diabatic coupling between the $(n, 3s)$ and (n, π^*) states, cf. Fig. 7.12. The following discussion is based on the simulations excluding all fields. The effective diabatic coupling is calculated from Eq. (4.36) at the center of the wavepacket in the $(n, 3s)$ state. As mentioned previously, the modes that couple the $(n, 3s)$ and (n, π^*) states in cyclobutanone are Franck-Condon active which is directly reflected in the non-zero value of the effective coupling at time zero. In cyclopentanone on the other hand, the effective coupling is very small close to time zero as the coupling modes are not Franck-Condon active. As the system evolves, internal vibrational energy redistribution (IVR) in the $(n, 3s)$ state induced by the third-order intra-state coupling terms of the VCHAM will transfer energy to the coupling modes on a 15–20 ps timescale, cf. Fig. 7.13b. IVR is thereby responsible for the gradual increase in the effective coupling observed in Fig. 7.12b.

Fourier transformation of the effective coupling reveals which modes are important in modulating the population transfer, cf. Fig. 7.14. In cyclobutanone, the dominating contributions are linear in the ring-puckering (peak at $\sim$0.03 eV) and C=O out-of-plane bend ($\sim$0.07 eV) modes. The symmetric C-CO-C stretch ($\sim$0.11 eV) and C=O stretch ($\sim$0.19 eV) have a smaller effect on the coupling. In cyclopentanone, the situation is markedly different. On the short timescale, the coupling is mostly modulated by the breathing motion of the wavepacket in the asymmetric C-CO-C stretch mode ($\sim$0.25 eV). On a longer timescale, the coupling is mostly modulated by combined motion in the ring-puckering and the asymmetric C-CO-C stretch modes ($\sim$0.05, $\sim$0.10, and $\sim$0.15 eV). Although cyclobutanone and cyclopentanone exhibit significantly different behavior, the involvement of low-frequency modes in modulating the effective coupling is a common trait as expected from experiment.

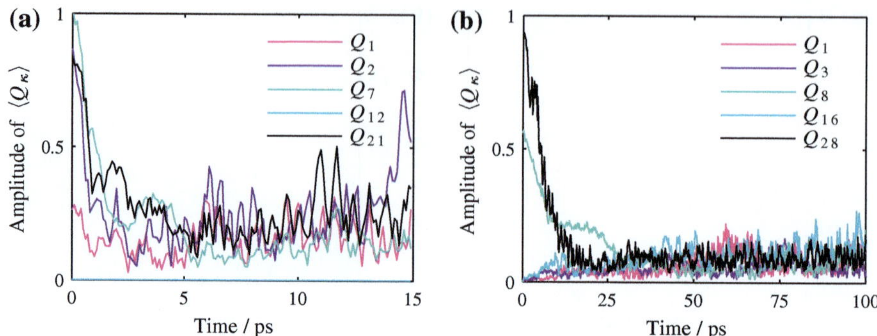

Fig. 7.13 Amplitudes of the oscillations of the coordinate expectation values in the $(n, 3s)$ state. Adapted with permission from Kuhlman et al. [24]. Copyright 2012, American Institute of Physics. (**a**) Cyclobutanone. (**b**) Cyclopentanone

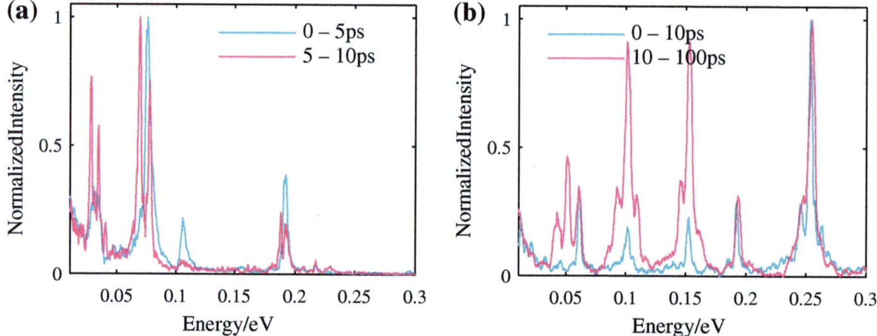

Fig. 7.14 Fourier transform of the effective diabatic coupling between the (n, π^*) and $(n, 3s)$ states. A low-pass filter has been applied to filter out noise from the spectrum. (**a**) Cyclobutanone. (**b**) Cyclopentanone

However, it should be noted that the very low vibrational frequencies observed in the experiments are not reproduced by the calculations.

The difference of timescales covers two significantly different dynamical pictures which to a large extent hinge on the electronic state symmetries of the two molecules: one direct and one indirect mechanism of population transfer. In cyclobutanone, the direct picture is prominent which results from initial motion in the reactive coupling modes. For cyclopentanone, the reactive coupling modes are not activated initially, and the indirect picture is most prominent. In this picture, the energy is initially deposited in non-reactive modes and a bottleneck in phase space results in a large component of delayed population decay as IVR is necessary for mediating the transfer of energy to the reactive modes. This energy transfer occurs on a 15–20 ps timescale, and as the population transfer occurs on a longer timescale ergodic behavior is approached.

7.2.3 *Time-Resolved Photoelectron Spectra*

We have calculated the time-resolved photoelectron spectra for cyclobutanone and cyclopentanone. For cyclobutanone, the calculated spectra at two different values of the detuning are compared to the experimental spectrum in Fig. 7.15. The spectra are constructed from separate calculations each employing a different value of Δt, i.e. the pump-probe delay, with a 25 fs spacing. The dependence of the timescale of population transfer on the detuning is reflected in the energy-integrated spectra in Fig. 7.16. This figure also exhibits the correlation between the $(n, 3s)$ population and the intensity of the photoelectron band assumed in the interpretation of the experimental data. The good correspondence between calculated and experimental spectra allow us to believe that the essential features of the dynamics have been captured by the reduced-dimensional model employed. However, we do not observe the spectral oscillation of the photoelectron band as was observed in experiment.

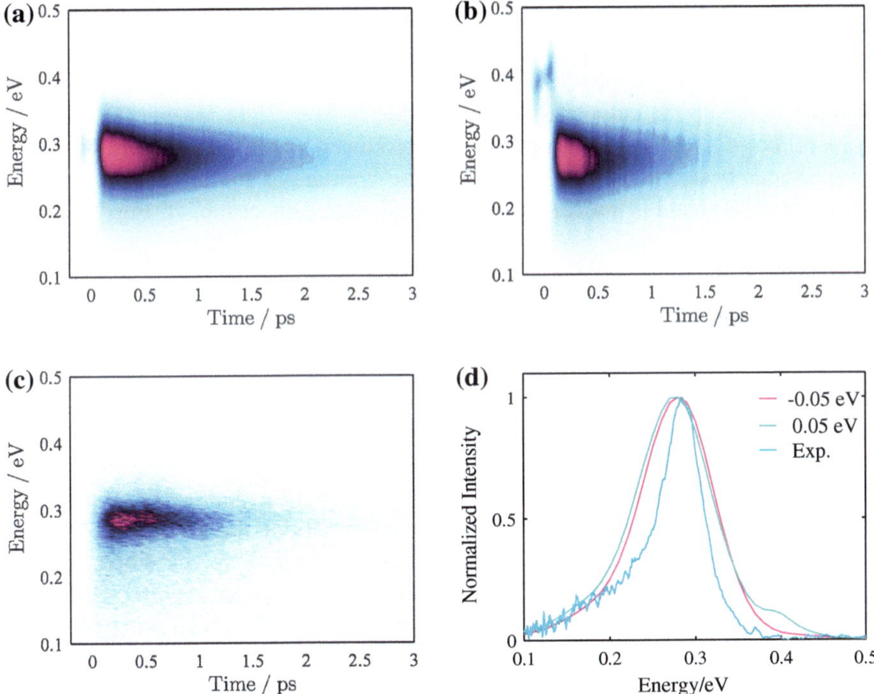

Fig. 7.15 (**a**)–(**c**) Experimental and calculated, for two different values of the detuning, time-resolved photoelectron spectra of cyclobutanone. The calculated spectra have been shifted by -0.28 eV. (**d**) Time-integrated spectra. The experimental spectrum has been smoothed by a moving average filter with a resolution of 0.01 eV corresponding to the energy resolution of the calculated spectra

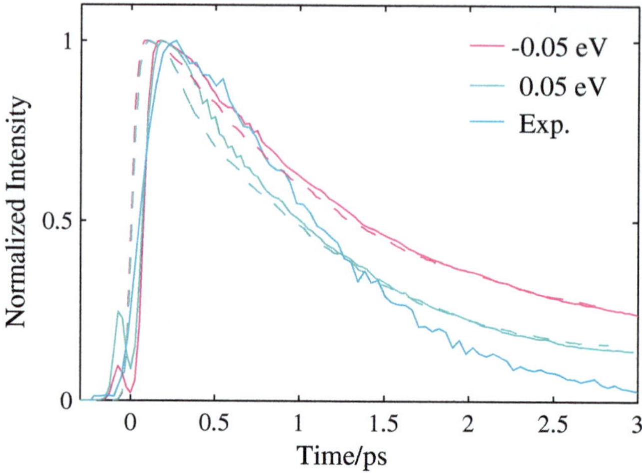

Fig. 7.16 Experimental and calculated, for two different values of the detuning, energy-integrated spectra of the $(n, 3s)$ photoelectron band of cyclobutanone. The *dashed lines* indicate the $(n, 3s)$ population from the simulations

For cyclopentanone, the spacing in the value of Δt between the separate calculations making up the calculated spectrum ranges from 25 to 500 fs.[1] The spectral width of the $(n, 3s)$ photoelectron band is reproduced by the calculations, but due to the significantly slower population decay in the simulations compared to experiment, the decay of the band is not captured by the calculated spectrum, cf. Fig. 7.17.

7.3 Conclusion

Using a set of seven cycloketones as model system, we have revealed several salient features of the complex process of internal conversion. Most conclusions derive from the observation that the process leading to transition from one electronic state to another, and, thereby, to the transformation of electronic energy into vibrational energy, is inherently localized—only one or a few vibrational modes factor in. As observed experimentally, merely by small structural variations, the vibrational frequency and the energy available in the upper excited state can be affected thereby tuning the rate of internal conversion over a range of more than an order of magnitude. A lower frequency and a larger available energy result in a faster process as the molecule can reach a configurational space in closer proximity of the crossing point between the excited states. The total density of vibrational states plays a smaller secondary role as an increase in this only leads to a very slight increase in the overall rate. In contrast to the standard energy gap laws that neglect the nuclear dependence

---

[1] For $\Delta t \leq 15$ ps. For $\Delta t > 15$ ps, the spacing is 5 ps.

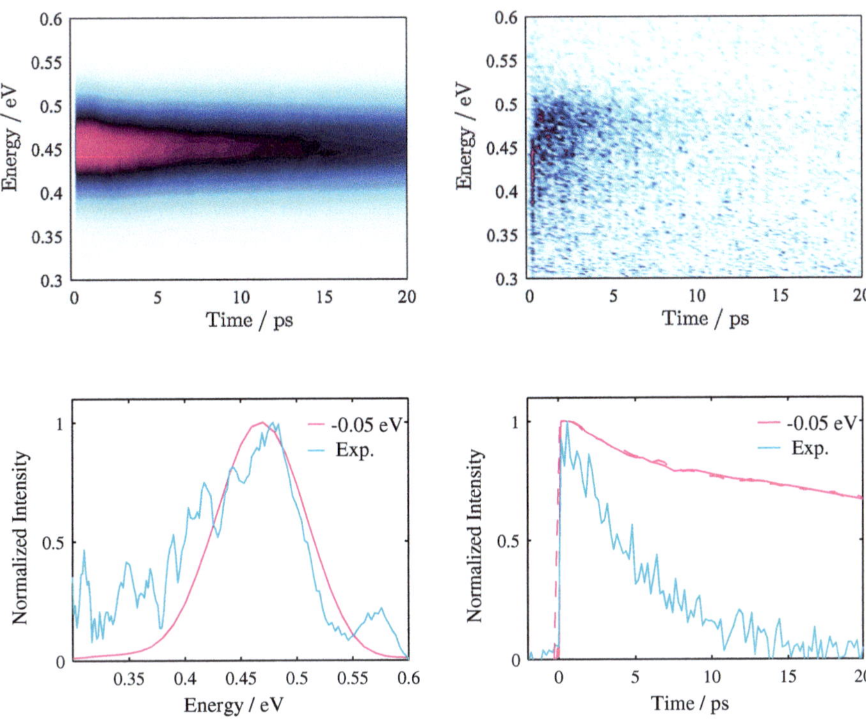

Fig. 7.17 (**a**), (**b**) Experimental and calculated time-resolved photoelectron spectra of cyclopentanone. The calculated spectrum has been shifted by -0.23 eV. (**c**) Time-integrated spectra. The experimental spectrum has been smoothed by a moving average filter with a resolution of 0.01 eV corresponding to the energy resolution of the calculated spectrum. (**d**) Energy-integrated spectra of the $(n, 3s)$ photoelectron band. The *dashed lines* indicates the $(n, 3s)$ population from the simulation

on the electronic coupling [16, 17], our results clearly show the effect of coherent nuclear motion on these matrix elements.

The effect of nuclear motion on the electronic coupling is also exhibited in the wavepacket simulations on cyclobutanone and cyclopentanone. In these calculations, it is noticeable that low-frequency modes play a central role in both molecules as also conjectured on the basis of the experimental results. However, for cyclopentanone the C-CO-C asymmetric stretch mode also factors in to a significant degree in the simulations. Whereas the time-resolved photoelectron spectrum of cyclobutanone can be very well reproduced from the wavepacket simulations, this is not the case for cyclopentanone. This discrepancy is likely due to the restriction of the VCHAM to five modes. Other modes not included could enhance the rate of IVR into one specific or several reactive coupling modes resulting in a faster rate of population transfer for cyclopentanone in closer agreement with experiment.

References

1. N.J. Turro, V. Ramamurthy, J.C. Scaiano, *Modern Molecular Photochemistry of Organic Molecules* (University Science Books, Sausalito, 2010)
2. E.W.-G. Diau, C. Kötting, A.H. Zewail, ChemPhysChem **2**, 273–293 (2001)
3. E.W.-G. Diau, C. Kötting, A.H. Zewail, ChemPhysChem **2**, 294–309 (2001)
4. E.W.-G. Diau, C. Kötting, T.I. Sølling, A.H. Zewail, ChemPhysChem **3**, 57–78 (2002)
5. T.I. Sølling, E.W.-G. Diau, C. Kötting, S. De Feyter, A.H. Zewail, ChemPhysChem **3**, 79–97 (2002)
6. M.A. El-Sayed, J. Chem. Phys. **38**, 2834–2838 (1963)
7. S.K. Lower, M.A. El-Sayed, Chem. Rev. **66**, 199–241 (1966)
8. M.A. El-Sayed, Acc. Chem. Res. **1**, 8–16 (1968)
9. M. O'Sullivan, A.C. Testa, J. Phys. Chem. **77**, 1830–1833 (1973)
10. T.J. Cornish, T. Baer, J. Am. Chem. Soc. **109**, 6915–6920 (1987)
11. T.J. Cornish, T. Baer, J. Am. Chem. Soc. **110**, 3099–3106 (1988)
12. L. O'Toole, P. Brint, C. Kosmidis, G. Boulakis, P. Tsekeris, J. Chem. Soc. Faraday Trans. **87**, 3343–3351 (1991)
13. L. O'Toole, P. Brint, C. Kosmidis, G. Boulakis, A. Bolovinos, J. Chem. Soc. Faraday Trans. **88**, 1237–1243 (1992)
14. T.S. Kuhlman, T.I. Sølling, K.B. Møller, ChemPhysChem **13**, 820–827 (2012)
15. T.S. Kuhlman, M. Pittelkow, T.I. Sølling, K.B. Møller, Angew. Chem. Int. Ed. **52**, 2247–2250 (2013)
16. M. Bixon, J. Jortner, J. Chem. Phys. **48**, 715–726 (1968)
17. R. Englman, J. Jortner, Mol. Phys. **18**, 145–164 (1970)
18. T.R. Borgers, H.L. Strauss, J. Chem. Phys. **45**, 947–955 (1966)
19. D.C. Moule, J. Chem. Phys. **64**, 3161–3168 (1976)
20. K. Tamagawa, R.L. Hilderbrandt, J. Phys. Chem. **87**, 5508–5516 (1983)
21. T. Beyer, D.F. Swinehart, Commun. ACM **16**, 379 (1973)
22. R.A. Marcus, J. Chem. Phys. **20**, 359–364 (1952)
23. K.B. Møller, A.H. Zewail, Femtosecond activation of reactions: the concepts of nonergodic behavior and reduced-space dynamics, in *Essays in Contemporary Chemistry: From Molecular Structure towards Biology*, Chap. 5, ed. by G. Quinkert, M.V. Kisakürek (Verlag Helvetica Chimica Acta, Zürich, and Wiley-VCH, Weinheim, 2001), pp. 157–188
24. T.S. Kuhlman, S.P.A. Sauer, T.I. Sølling, K.B. Møller, J. Chem. Phys. **137**, 22A522 (2012)
25. T.H. Dunning, J. Chem. Phys. **90**, 1007–1023 (1989)
26. R. Kosloff, H. Tal-Ezer, Chem. Phys. Lett. **127**, 223–230 (1986)

Chapter 8
The Cyclopentadienes

Parts of the results described in this chapter have been published in Kuhlman et al. [1]. Some text passages and/or figures and tables in this chapter have been reproduced by permission of The Royal Society of Chemistry.

Molecules possessing π electrons play a central role in organic photochemistry and photophysics [2]. In the previous chapter, we encountered examples of such molecules all containing the carbonyl chromophore. In this chapter, we will focus on non-aromatic molecules containing carbon–carbon double bonds, i.e. alkenes. Alkenes participate in a plethora of reactions induced by light such as photoisomerization [3–10], electrocyclic ring opening and closing [11–22], sigmatropic rearrangement [23–25], and cycloaddition [26, 27].

The rich set of photoinduced phenomena exhibited by alkenes is a consequence of the complex nature of the excited states of π electron systems and conjugated systems in particular. This is evident even in the simplest case of ethylene where the lowest excited state can have different character at the Franck-Condon point and twisted geometries [28]. For polyenes, the picture is even more complicated due to the presence of a low-lying electronic state with a large doubly excited character [29]. Due to the optically dark nature of this state it often eludes direct observation, however, it is well established that it plays a significant role in the photochemistry of polyenes with more than three conjugated double bonds [29].

In between the distinct ethylene and the longer polyenes, the role of the doubly excited state in the dienes is more subtle. In the case of molecules with the two double bonds in a *trans*-configuration with respect to the connecting single bond, i.e. *s-trans*-dienes, cf. Fig. 8.1, numerous studies have investigated the state ordering of the bright (π, π^*) state and the doubly excited state [30–36]. Both experimental and theoretical studies suggest that the doubly excited state plays a significant role in the initial dynamics following excitation to the bright state giving the *s-trans*-dienes some of the characteristics of the longer polyenes. However, the longer time dynamics resemble those of ethylene [37]. In the case of *s-cis*-dienes and cyclopentadienes in particular, numerous experimental studies have investigated the low-lying valence states [38–41], but only few have alluded to a discussion of the spectral position

T. S. Kuhlman, *The Non-Ergodic Nature of Internal Conversion*, Springer Theses, DOI: 10.1007/978-3-319-00386-3_8, © Springer International Publishing Switzerland 2013

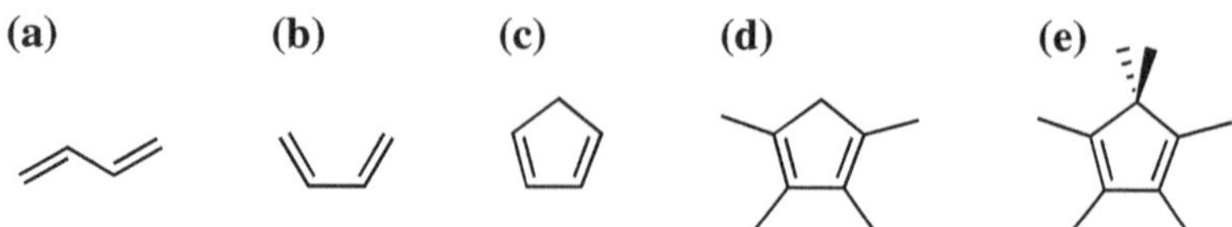

Fig. 8.1 Structures of (**a**) *s-trans*-butadiene, (**b**) *s-cis*-butadiene, (**c**) cyclopentadiene (CPD), (**d**) 1,2,3,4-tetramethylcyclopentadiene (Me$_4$CPD), and (**e**) hexamethylcyclopentadiene (Me$_6$CPD)

of the doubly excited state [42, 43]. Despite the inability to directly locate this dark state by absorption spectroscopy, several time-resolved mass spectrometry and photoelectron spectroscopy (TRPES) studies have invoked the state to explain the dynamics observed upon excitation to the bright (π, π^*) state [25, 44, 45].

In the following, we attempt to unravel the dynamics of three cyclopentadienes, cyclopentadiene (CPD), 1,2,3,4-tetramethylcyclopentadiene (Me$_4$CPD), and hexamethylcyclopentadiene (Me$_4$CPD), following excitation to the bright (π, π^*) state by use of Ab Initio Multiple Spawning (AIMS). In all calculations, the electronic structure is calculated for cyclopentadiene, and the methylated species are approximately treated in the dynamics simulations by setting the mass of the substituted hydrogens to that of a methyl group. The observed dynamics are discussed in light of the dynamics of ethylene and other polyenes in particular the similar *s-trans*-butadiene. Finally, we relate the results to experimental data from TRPES by direct calculation of spectra by use of the perturbative method described in Sect. 5.1.

8.1 Electronic Structure

The two lowest valence excited states in cyclopentadiene are dominated by $\pi \rightarrow \pi^*$ and $\pi^2 \rightarrow (\pi^*)^2$ promotions at the Franck-Condon geometry. These states are termed V_1 and V_2 in Mulliken notation,[1] and we will retain these labels as diabatic state labels referring to the electronic character whereas S$_1$ and S$_2$ will be strictly adiabatic labels [46]. The doubly excited character of V_2 can be understood as arising from excitation of each ethylene unit to its lowest triplet state but spin-coupled to an overall singlet [47]. Our MS-MR-CASPT2/6-31G(d,p) [48, 49] calculations use an active space of four electrons distributed in four orbitals, the two π and the two π^* orbitals, with state averaging over the ground and the lowest two excited states. A level shift of 0.2 Hartrees was employed. Using this method, the vertical transition energy to the S$_1$ state is 5.46 eV in very good agreement with the best estimate of 5.43 ± 0.05 eV found from a combination of high-level theoretical methods and spectroscopic simulations [50] and in the range 5.19–6.46 eV determined using

---

[1] Strictly, the V labels of Mulliken refer to singly excited configurations, however, the V_2 state mixes with a higher lying doubly excited configuration to attain partial doubly excited character, and we use the V_2 label to refer to this state and character.

various high-level electronic structure methods [51–56]. The calculated vertical value is slightly different from the spectral position of the band maximum of 5.17–5.33 eV found from experiment [38–40, 42, 45]. Similarly, we find the vertical excitation energy to the S_2 state to be 6.51 eV, a value which falls in between previous calculated values of 6.31–7.05 eV [51, 52, 54–56] and slightly above the value of 6.2 eV suggested on the basis of experimental results [42]. This state possesses a large doubly excited character of $\sim$50 %.

Resonance Raman depolarization ratios suggest that the minimum of V_2 lies below that of V_1, and a conical intersection connecting these two states is thus expected [43]. We indeed locate a minimum energy conical intersection (MECI) between S_2 and S_1 which furthermore corresponds to the minimum energy configuration on the S_2 potential energy surface located in this work. This S_2S_1 MECI is akin to the crossing of two non-interacting diabatic states—one similar in character to V_2 and one similar to V_1. We have also located three MECIs connecting S_1 with the ground state on the same intersection seam. Two of these MECIs result primarily from twisting of a single double bond and are therefore termed ethylene-like and referred to as $eth1$ and $eth2$. The last MECI results from a disrotatory mechanism where both double bonds twist to some degree and is thus termed dis. The $eth1$ MECI is the lowest energy configuration on the S_1 potential energy surface located in this work. The three S_1S_0 MECIs correspond to the crossing between a state somewhat similar in character to V_1 and the ground state—the first being ionic in character with charge separation between the two carbons of the double bonds (larger charge separation for the most twisted double bond) and the second being of a more diradicaloid character. However, the $\sim$25 % doubly excited character of the first state is an order of magnitude larger than what is found for the S_1 state at the Franck-Condon geometry. The geometries of the four MECIs are depicted in Fig. 8.2.

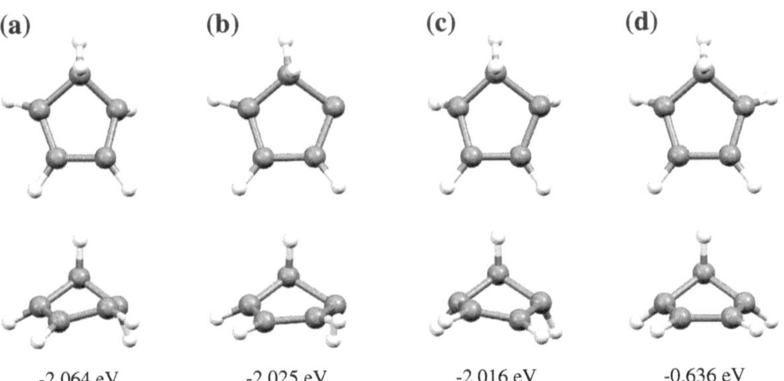

Fig. 8.2 Geometries of the four MECIs viewed from two different angles along with the relative energy of the S_1 state with respect to the energy at the Franck-Condon point. All four MECIs are energetically accessible after excitation to the S_1 state [1]—reproduced by permission of The Royal Society of Chemistry. (**a**) $eth1$ MECI, (**b**) $eth2$ MECI, (**c**) dis MECI, (**d**) S_2S_1 MECI

Table 8.1 Details of the AIMS simulations for CPD/Me$_4$CPD/Me$_6$CPD

Initial conditions	0 K Wigner distribution of the harmonic ground vibrational state [57]
Initial trajectories	40
Total trajectories	179/212/205
Timestep/fs	0.387 (adaptive)
Total time/fs	190/240/290
Integrator CM	Velocity-Verlet
Integrator QM	Adaptive second-order Runge-Kutta

As indicated in Fig. 8.2, all four MECIs are energetically accessible after excitation to the S_1 state, and there are no barriers on the path from the Franck-Condon geometry to either MECI (cf. Fig. B.1 in Appendix B). Furthermore, all MECIs have a peaked geometry in the branching space (cf. Fig. B.2 in Appndix B) which leads us to expect very fast non-adiabatic transitions. On the other hand, it is not a priori possible to determine whether the molecule visits the part of the S_1 potential energy surface that can be directly associated with the V_2 state and what role this latter state plays in the dynamics following excitation.

8.2 Dynamics Simulations

The details of the AIMS simulations are summarized in Table 8.1. The final time was chosen long enough to capture the essential dynamics on the excited states and to determine the timescales for population transfer. In cases where all population ($>$99 %) had been transferred to the ground state before the final time was reached, the calculation was stopped as our focus inhere is on the excited state dynamics and timescale of non-adiabatic transfer and not on possible thermal reactions taking place on the vibrationally hot ground electronic state.

8.2.1 Setting the Timescale of Population Transfer

The S_1 population decays in Fig. 8.3 reveal that all three molecules undergo full population transfer from the initially excited S_1 state back to S_0 on a sub 300 fs timescale. Very few spawning events from S_1 to S_2 are observed, and the total population transfer to S_2 is $<$0.1 % for all three molecules.

The onset of population decay is preceded by a delay period t_{in}—the so-called induction time [58, 59]. In the case of CPD, this period is $\sim$25 fs whereas it is $\sim$32 fs for Me$_4$CPD and $\sim$106 fs for Me$_6$CPD. To quantify the timescale of the subsequent population transfer, we determine the half-life $\tau_{1/2}$ defined as the time it takes the S_1 population to reach 0.5 following the induction period. As an alternative, we fit

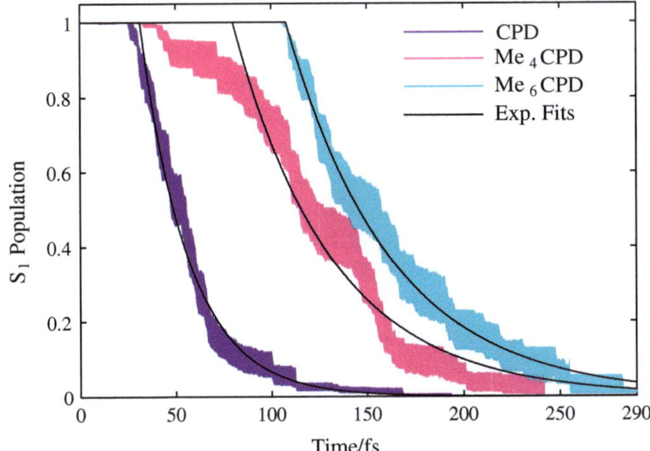

Fig. 8.3 Population of the S_1 state with standard deviations from bootstrapping indicated by the shaded regions. The fitted exponentials decays are also given

Table 8.2 Timescales for the S_1 population decay in CPD, Me$_4$CPD, and Me$_6$CPD (in fs) determined as either the induction time and the half-life or from a fitted exponential decay model

Molecule	t_{in}	$\tau_{1/2}$	t_0	τ
CPD	25	28	31 ± 2	25 ± 2
Me$_4$CPD	32	88	80 ± 7	52 ± 2
Me$_6$CPD	106	43	108 ± 3	55 ± 8

an exponential model

$$P(t) = \Theta(t - t_0) \exp[-(t - t_0)/\tau] + \Theta(t_0 - t) \qquad (8.1)$$

to yield a time t_0 similar in nature to the induction period and a time constant τ for the exponential decay. The sets of two timescales determined by these two methods are collected in Table 8.2. The results are in line with the observation from experiments of a slow-down of the dynamics upon methylation of CPD [45]. The following discussion will explore the background for these observed timescales and their differences.

8.2.2 Nuclear Dynamics

The initial dynamics following excitation to the S_1 state are in all three molecules characterized by significant in-plane nuclear motion as is common in conjugated molecules exemplified by *s-trans*-butadiene [37]. The promotion of an electron from a bonding π orbital to an antibonding π^* orbital leads to an elongation of the double bonds in conjunction with a contraction of the connecting single bond. This nuclear motion can be quantified by the bond alternation coordinate defined as the sum of the double bond lengths minus the length of the connecting single bond. The time evolution of the nuclear wavepacket density in the S_1 state for this coordinate is given in Fig. 8.4(left). A significant in-plane distortion is observed by the fast increase in the expectation value of the coordinate over the first $\sim$15 fs before oscillatory motion around the new equilibrium ensues after $\sim$25 fs. At this time, the in-plane motion towards the $S_1 S_0$ MECIs is completed. The value of the bond alternation coordinate at the *eth*1 MECI is given by the dashed lines in Fig. 8.4(left). The value of the coordinate for the other MECIs can be found in Table B.5 in Appendix B. It is evident that the in-plane motion is not the source of the different timescales observed for the population transfer between S_1 and S_0 in the three molecules.

In conjunction with in-plane nuclear motion, out-of-plane motion is also observed although this occurs on a longer timescale due to the low-frequency modes involved. Although the timescale for torsion in the carbon backbone differs between CPD and the methylated molecules, it does not distinguish Me_4CPD from Me_6CPD, cf. Fig. 8.4 (right). In addition to distortion in the ring structure, the MECI geometries depicted in Fig. 8.2 are also characterized by a significant twist of one (or both in the case of the *dis* MECI) of the doublebonds leading to the out-of-plane bend of the CX_2 group. X refers to a hydrogen for CPD and Me_4CPD and a methyl group in the case of Me_6CPD. The time evolution of the wavepacket density projected onto the double bond twist and CX_2 bend coordinates are given in Fig. 8.5. These coordinates are observed to be highly correlated. The timescale for out-of-plane motion in CPD is significantly shorter than that for the methylated species, and a clear distinction can also be observed between Me_4CPD and Me_6CPD. These timescale differences entail that the potential energy surface in the vicinity of the MECIs, where the coupling between S_1 and S_0 is large, is visited earlier by a larger part of the S_1 population for CPD compared to Me_4CPD and for Me_4CPD compared to Me_6CPD. This results in a decreasing rate of non-adiabatic transition in the order $CPD > Me_4CPD > Me_6CPD$.

From the spawning events, we can assign population transfer to one of the $S_1 S_0$ MECIs by using the spawning geometries as starting points for optimization of $S_1 S_0$ MECIs. This procedure reveals a bifurcation on the S_1 state for CPD with 71 % of the population transfer being attributable to the *eth*1 MECI and 27 % to *eth*2 MECI. A very small part of the population transfer, 2 %, can be assigned to the *dis* MECI. In the case of the methylated species, 73 and 98 % can be attributed to the *eth*1 MECI and 27 and 2 % to the *eth*2 MECI for Me_4CPD and Me_6CPD respectively. The slowdown of the dynamics in the methylated species apparently allows for a larger part of the S_1 population to reach the vicinity of the *eth*1 MECI, the lowest energy $S_1 S_0$ MECI, before population transfer proceeds.

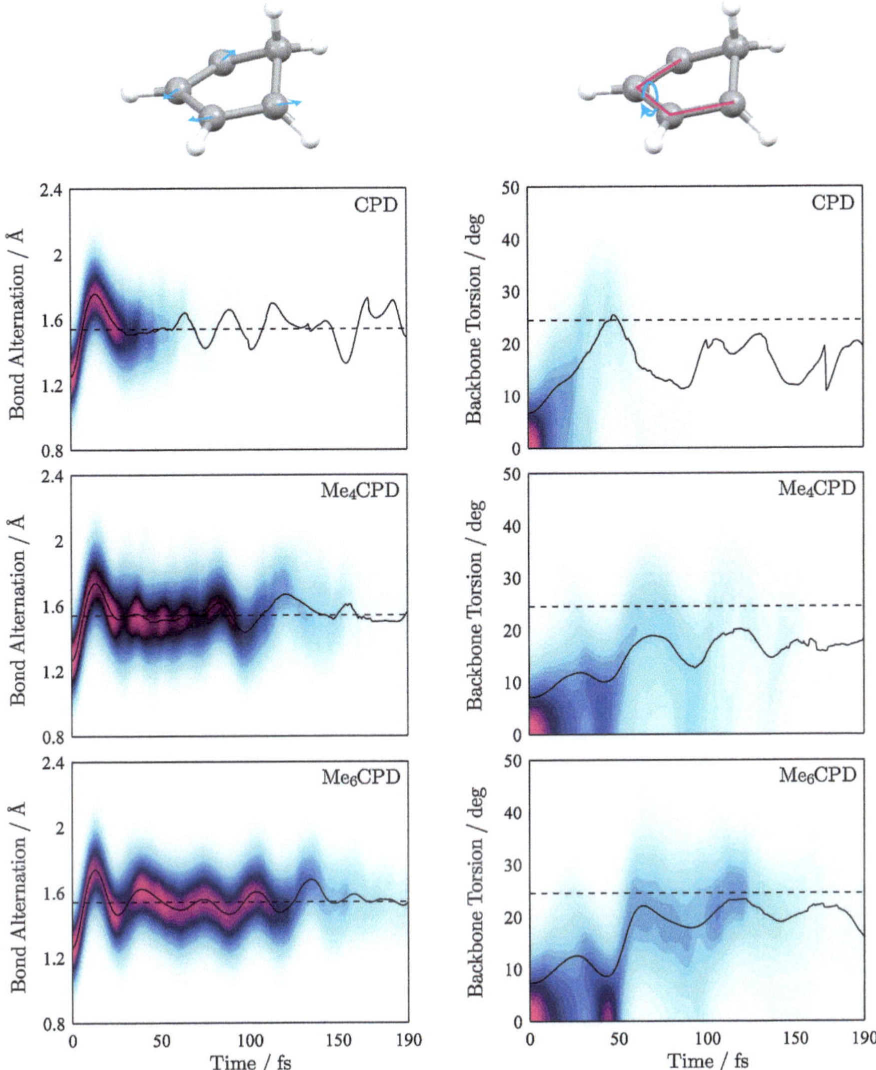

Fig. 8.4 *Left* Projection of the nuclear wavepacket density in the S_1 state onto the bond alternation coordinate and *right* projection onto the backbone torsion coordinate. The *solid black line* indicates the expectation value whereas the *dashed line* indicates the value of the coordinate at the *eth*1 MECI [1]—adapted by permission of The Royal Society of Chemistry

8.2.3 Electronic Dynamics

Having discussed the nuclear dynamics, we turn our attention to the electronic character of the states involved. For the *s-trans*-dienes, the two lowest excited states are

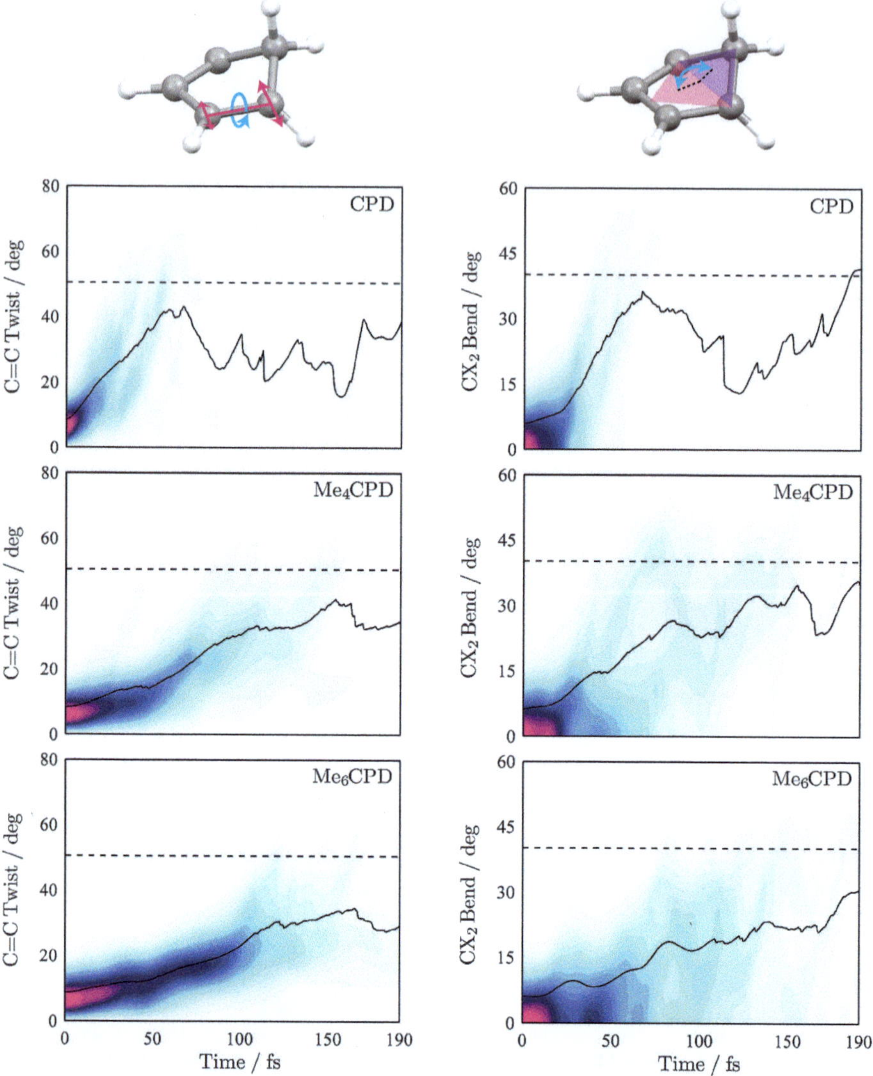

Fig. 8.5 *Left* Projection of the nuclear wavepacket density in the S_1 state onto the $C=C$ twist coordinate and *right* projection onto the CX_2 bend coordinate. The *solid black line* indicates the expectation value whereas the *dashed line* indicates the value of the coordinate at the *eth*1 MECI [1]—adapted by permission of The Royal Society of Chemistry

close in energy at the Franck-Condon geometry. In the case of *s-trans*-butadiene, an ultrafast exchange of electronic character between S_1 and S_2 takes place within the first 5 fs subsequent to excitation to the bright state [37]. This exchange of electronic character is reflected by a change in the transition dipole moment between S_0 and

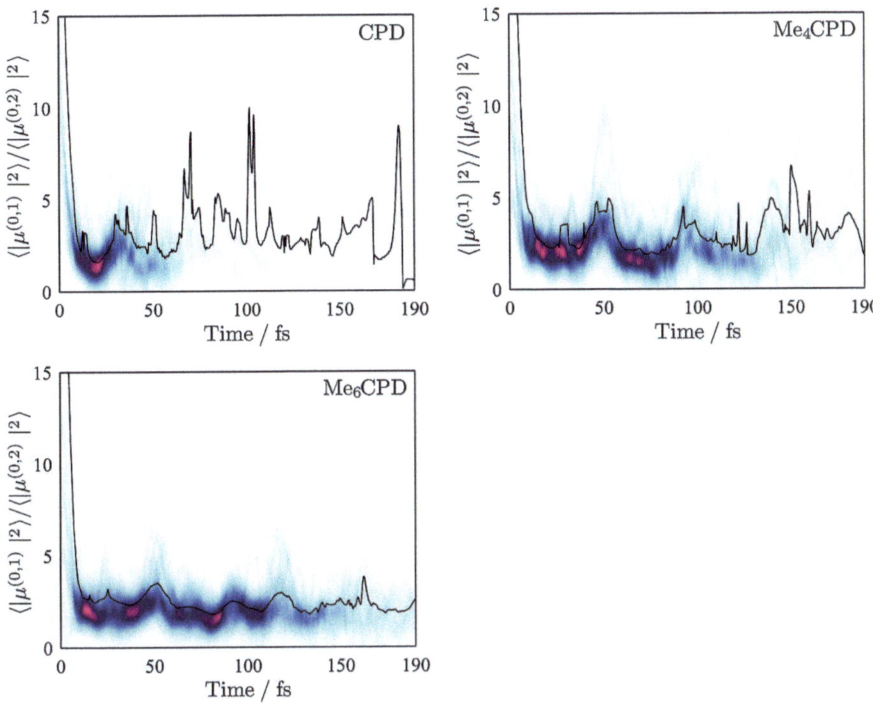

Fig. 8.6 Histogram of the average of the squares of the transition dipole moments between S_0 and S_1 and S_2. The histograms have been convoluted by a Gaussian of width 1.0 in the ordinate. The *black lines* indicate the average values [1]—reproduced by permission of The Royal Society of Chemistry

the excited states such that the initially bright S_1 state becomes dark, and the initially dark S_2 state becomes bright. The change of character is unambiguous for *s-trans*-butadiene. In the case of CPD (and similarly for Me_4CPD and Me_6CPD), the two lowest excited states are separated to a larger extent at the Franck-Condon geometry with a calculated energy splitting of 1.05 eV. At this geometry, the diabatic labels V_1 and V_2 were unambiguously assigned to the adiabatic states in Sect. 8.1 based on the electronic configurations. The transition dipole moments to S_0, calculated to be 2.81 D in the case of S_1 and 0.29 D in the case of S_2, also reflect this character of the states. Figure 8.6 exhibits the time evolution of the ratio of the squared transition dipole moments. For all three molecules, the ratio starts out > 10 (the value is ~ 100 at the Franck-Condon geometry), however, it drops within the first 10 fs to ~ 3. Thus, it is apparent that there is a mixing of the electronic character, and an unambiguous assignment to bright and dark (or equivalently to V_1 and V_2) of the two adiabatic states S_1 and S_2 is not possible at later times. The change in character of the adiabatic states is closely related to motion in the backbone torsion coordinate.

It has been observed for *s-trans*-butadiene that charge-transfer states play an essential role in the excited state dynamics [37]. In *s-trans*-butadiene, charge separation occurs on the S_1 state and is preceded by twisting of a single methylene unit akin to the twist of a single double bond in the cyclopentadienes. However, in the latter group of molecules, the torsional motion is frustrated due to the ring structure, and the twist does not reach the extremum corresponding to a complete 90° twist as is observed in *s-trans*-butadiene. As a consequence, a significantly smaller degree of charge separation across the double bonds is observed in the cyclopentadienes.

8.3 The Excited State Reaction Mechanism

Having established the nuclear dynamics leading to the non-adiabatic transition between S_1 and S_0 and the electronic character of the states involved, a complete picture of the excited state dynamics emerges. The dynamics are schematically summarized in Fig. 8.7. At the Franck-Condon geometry, the S_1 and S_2 states can clearly be identified as the bright V_1 and dark V_2 states respectively. Subsequent to excitation, initial nuclear motion primarily along in-plane modes, but to some extent also along out-of-plane modes, takes the molecules out of the Franck-Condon region in ~25 fs. As a consequence of this nuclear motion, the electronic character of the excited states

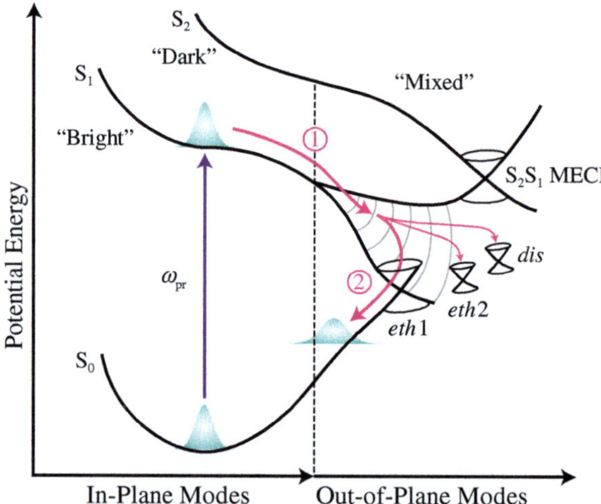

Fig. 8.7 Schematic summary of the excited state dynamics in the cyclopentadienes following excitation to S_1. ① The initial dynamics taking the molecules out of the Franck-Condon region primarily involve motion along in-plane modes, but in particular for CPD, out-of-plane motion also occurs. This motion leads to a mixing of the electronic state character of S_1 and S_2. ② Subsequent motion primarily along out-of-plane modes takes the molecules to the vicinity of the S_1S_0 MECIs wherefrom population transfer back to S_0 proceeds. A slight bifurcation between the two ethylene-like MECIs is observed, and in the case of CPD, a very small part of the population transfer can also be associated with the *dis* MECI [1]—adapted by permission of The Royal Society of Chemistry

mixes significantly, and an unambiguous assignment of diabatic labels to the adiabatic states S_1 and S_2 is no longer possible. After the initial nuclear motion, primarily out-of-plane motion takes the molecules towards the $S_1 S_0$ MECI geometries. This motion involves twisting of the double bonds very similar to what is observed for *s-trans*-butadiene and the smaller ethylene. Although the two ethylene-like $S_1 S_0$ MECIs, to which most population transfer can be assigned, primarily result from twisting of only one double bond, the spawning geometries reveal a slight disrotatory mechanism where some twisting also occurs around the other double bond. For the spawning geometries, the difference in twist of the two double bonds falls somewhere in between that for the ethylene-like $S_1 S_0$ MECIs and the *dis* MECI. The out-of-plane motion is slowed down in Me_4CPD and Me_6CPD compared to CPD due to the inertia of the methyl substituents. This targeted substitution and the consequential slowdown in the population transfer truly exhibits the localized nature of the dynamics—the double bond twist is the primary degree of freedom of importance.

8.4 Time-Resolved Photoelectron Spectra

On the basis of the dynamics simulations, two timescales of importance have been identified—one timescale during which nuclear motion takes place only on the initially excited state with the molecule moving away from the Franck-Condon region and one timescale for the non-adiabatic transfer back to the ground state. By use of TRPES, two timescales have also been identified in experimental data [45]. To be able to make a direct comparison to experiment, time-resolved photoelectron spectra were calculated on the basis of the simulated dynamics. The two lowest cationic states were included in the calculations, and a Gaussian window function of width $0.10\,eV$ was used in conjunction with $\tau_{xc} = 160\,fs$ and a probe energy of $\omega_{pr} = 3.87\,eV$ unless otherwise specified. The shifts employed were $\Delta_{CPD}^{(S_1,D_0)} = -0.46\,eV$ and $\Delta_{Me_4CPD}^{(S_1,D_0)} = 0.39$.

Figure 8.8 shows a comparison between the calculated and the experimental spectrum of CPD. The calculated spectrum is a sum of the spectrum calculated for ionization by one probe photon and that for ionization by a photon of twice the frequency of the probe photon to approximately treat two-probe-photon ionization. For the calculation of the latter spectrum, the Condon approximation was invoked. The relative maximum intensities of the one-photon and two-photon spectra were fixed at 20 following experimental findings [45]. Both the calculated and the experimental spectrum exhibit a band $<0.5\,eV$ centered near time zero due to one-photon ionization and a broad delayed band due to two-photon ionization. Thus, it is evident that the present simulation is able to qualitatively reproduce the experimental spectrum, however, the timescales from the simulation are shorter than those obtained from the experimental spectrum of 39 and 51 fs [45].

Both bands of the calculated spectrum are observed to originate from ionization out of the S_1 state. The disappearance of the low-energy one-photon band is a consequence of a fast increase in ionization potential from S_1 to D_0, the ground state of

Fig. 8.8 Time-resolved photoelectron spectra of CPD. The experimental spectrum, based on data published in Ref. [45] and reproduced with permission, was obtained using $\omega_{pu} = 5.19\,\text{eV}$, $\omega_{pr} = 3.87\,\text{eV}$, and $\tau_{xc} = 160\,\text{fs}$. The spectra have been multiplied by a factor of 20 in the region above 1.0 eV [1]—Adapted by permission of The Royal Society of Chemistry. (**a**) Calculated spectrum. (**b**) Experimental spectrum

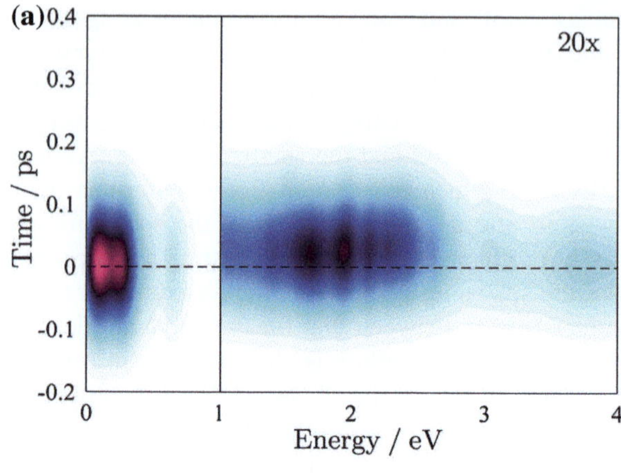

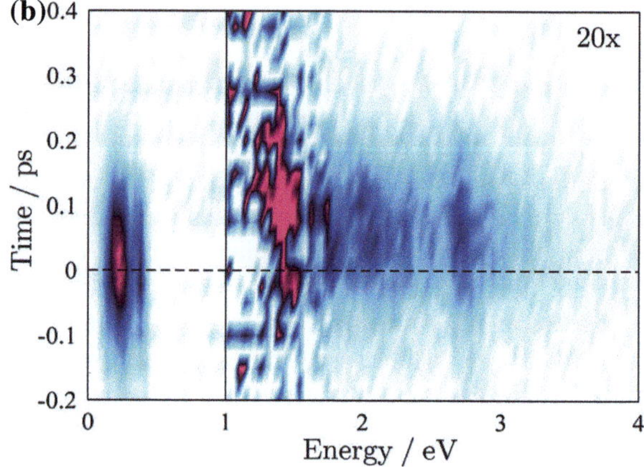

the cation, when the molecule leaves the Franck-Condon region and slides down the potential energy surface. It is thus the energetic factor of Eq. (5.14) that leads to the decay of the low-energy band by effectively closing the one-photon probe window. Through two-photon ionization, another probe window is open further down the potential energy surface resulting in the band centered at a kinetic energy of 1.9 eV. This window stays open longer than the one-photon window until population decay back to the ground state finally leads to the decay of the band.

To better follow the dynamics, an ideal experiment could be constructed by using shorter pulses and a higher frequency probe. One possibility is to use $\omega_{pr} = 6.10\,\text{eV}$ as the absorption coefficient for CPD at this energy is relatively low such as to avoid probe-induced dynamics in an experiment [42]. The calculated spectrum of CPD using this probe energy with $\tau_{xc} = 20\,\text{fs}$ is exhibited in Fig. 8.9. The appearance of the spectrum can be rationalized using the schematic in Fig. 8.10. The relatively narrow wavepacket formed in the S_1 state by the excitation process gives rise to an

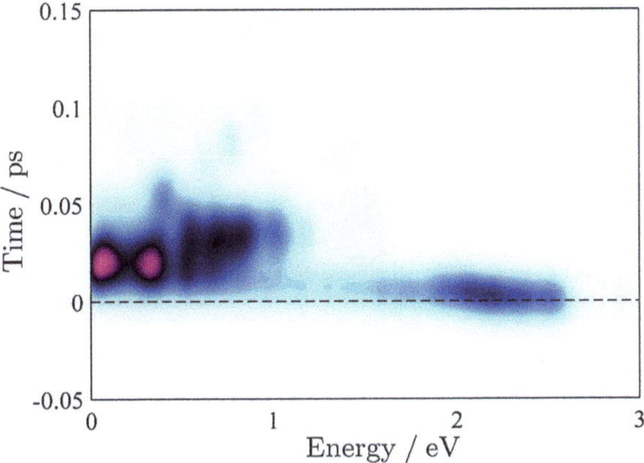

Fig. 8.9 Time-resolved photoelectron spectrum of CPD. The spectrum was calculated using $\tau_{xc} = 20\,\mathrm{fs}$ and $\omega_{pr} = 6.10\,\mathrm{eV}$

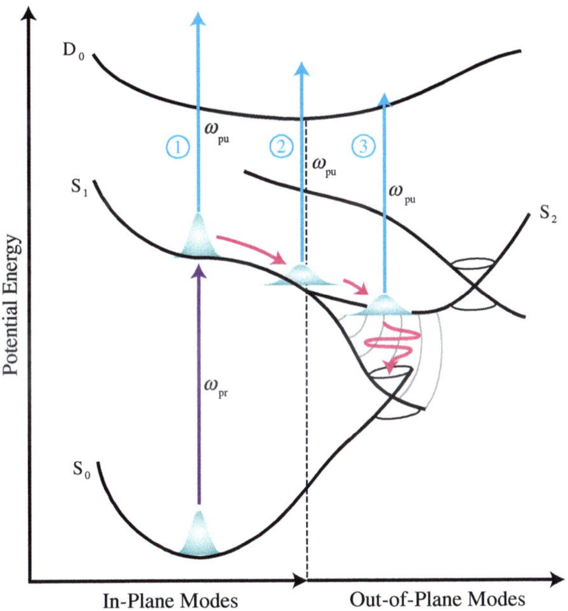

Fig. 8.10 Final scheme for qualitatively explaining the appearance of the time-resolved photoelectron spectra. ① The narrow wavepacket subsequent to excitation gives rise to an intense band at high energy, ② the spreading wavepacket on its way down the potential energy surface gives rise to a diffuse band at intermediate energies, and ③ recurrences of the wavepacket near the bottom of the S_1 potential energy surface gives rise to an intense band at low energies. Depending on the probe energy and the ionization potential of the molecule, the ionization process can be cut-off as the wavepacket moves down the S_1 potential energy surface

Fig. 8.11 Time-resolved photoelectron spectra of Me$_4$CPD. The experimental spectrum, based on data published in Ref. [45] and reproduced with permission, was obtained using $\omega_{pu} = 5.19$ eV, $\omega_{pr} = 3.87$ eV, and $\tau_{xc} = 160$ fs. (a) Calculated. (b) Experimental

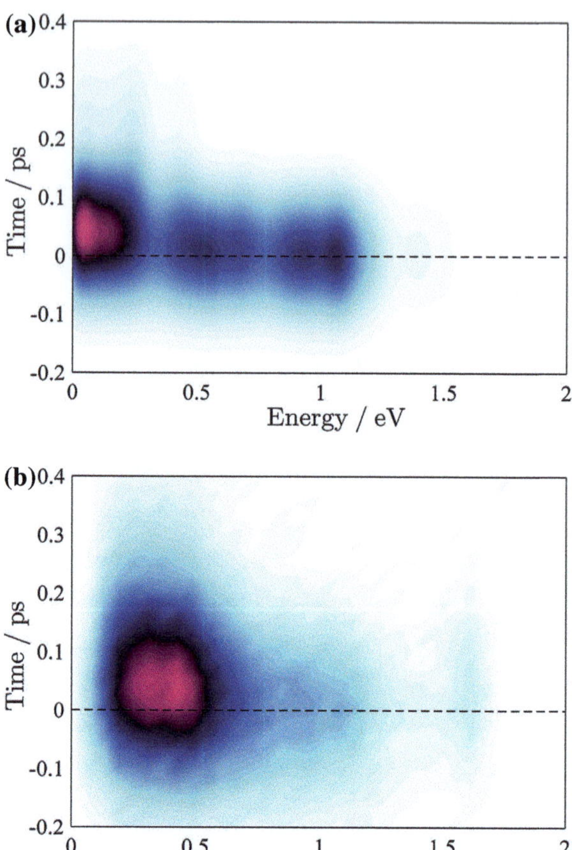

intense band between 2.0 and 2.5 eV. On its way down the S$_1$ potential energy surface, the wavepacket spreads somewhat giving rise to a more diffuse band between 1.0 and 2.0 eV the center of which red-shifts with time. At the bottom of the S$_1$ potential energy surface near the S$_1$S$_0$ MECIs, recurrences of the wavepacket give rise to an intense band between 0.0 and 1.0 eV. In this band, reminiscence of vibrational structure can be observed reflecting the coherent out-of-plane motion of the molecule.

Figure 8.11 shows a comparison between the calculated and experimental spectrum of Me$_4$CPD. The features of the calculated spectrum can qualitatively be explained by the model in Fig. 8.10 similar to the previous spectrum for CPD. The correspondence between the calculated and experimental spectrum is not as good as in the case of CPD which could be expected as the calculated spectrum is based on the electronic structure of CPD and not Me$_4$CPD. In particular, the initial peak at the onset of excitation is less intense in the experimental spectrum. The discrepancy between the spectra at energies <0.15 eV is most likely due to the insensitivity of the experiment to very low-energy electrons. Similar to the case of CPD, the combined timescales from the simulation are slightly shorter than those obtained from the experimental spectrum of 68 and 76 fs [45].

8.5 Conclusion

The dynamics of the cyclopentadienes are truly in between those of ethylene and the poylenes. The initial dynamics following excitation to the S_1 state are similar to the polyenes with a large chance in the bond alternation coordinate in the direction of the gradient at the Franck-Condon geometry. At longer times, symmetry breaking due to out-of-plane motion leads to localization of the dynamics on primarily one of the ethylene units as the molecule moves towards the ethylene-like conical intersections connecting S_1 with S_0. This out-of-plane motion is induced by a negative curvature along out-of-plane modes at the Franck-Condon geometry. The non-ergodicity of the dynamics is exhibited by the significant slowdown of the non-adiabatic population transfer upon methyl substitution which primarily affects the out-of-plane motion. As the methyl groups are only approximately treated by changing the mass of the pertinent hydrogens, the slowdown observed in the dynamics is truly a kinematic effect and cannot be explained by a statistical theory or a change in the electronic structure.

Although dynamics on the doubly excited V_2 state are not directly observed in the simulations, this state does play a role in the dynamics. The mixing of the electronic character of the V_1 and V_2 states in the adiabatic S_1 and S_2 states, i.e. due to the coupling between the S_1 and S_2 states, results in a low-energy seam of conical intersections between the S_1 and S_0 states where the former state has partially doubly excited character. V_2 is thus implicitly involved in the ultrafast dynamics observed in the cyclopentadienes.

The coupled electronic and nuclear dynamics in the cyclopentadienes is exhibited in the time-resolved photoelectron spectra. However, it is also clear that features in the spectra which at first glance are attributed to non-adiabatic dynamics are indeed due to a closing probe window. By using a higher frequency probe, one should be able to follow the full dynamics on the S_1 potential energy surface and more directly determine a timescale for the $S_1 \rightarrow S_0$ transition from experiment.

References

1. T.S. Kuhlman, W.J. Glover, T. Mori, K.B. Møller, T.J. Martínez, Faraday Discuss. **157**, 193–212 (2012)
2. N.J. Turro, V. Ramamurthy, J.C. Scaiano, *Modern Molecular Photochemistry of Organic Molecules* (University Science Books, Sausalito, 2010)
3. D.H. Waldeck, Chem. Rev. **91**, 415–436 (1991)
4. B.E. Kohler, Chem. Rev. **93**, 41–54 (1993)

5. M. Ben-Nun, T.J. Martínez, Chem. Phys. Lett. **298**, 57–65 (1998)
6. J. Quenneville, T.J. Martínez, J. Phys. Chem. A **107**, 829–837 (2003)
7. W. Fuß, C. Kosmidis, W.E. Schmid, S.A. Trushin, Angew. Chem. Int. Ed. **43**, 4178–4182 (2004)
8. C. Dugave, L. Demange, Chem. Rev. **103**, 2475–2532 (2003)
9. B.G. Levine, T.J. Martínez, Annu. Rev. Phys. Chem. **58**, 613–634 (2007)
10. M. Barbatti, M. Ruckenbauer, J.J. Szymczak, A.J.A. Aquino, H. Lischka, Phys. Chem. Chem. Phys. **10**, 482–494 (2008)
11. R. Hoffmann, R.B. Woodward, Acc. Chem. Res. **1**, 17–22 (1968)
12. R.B. Woodward, R. Hoffmann, Angew. Chem. Int. Ed. **8**, 781–853 (1969)
13. W. Fuß, W.E. Schmid, S.A. Trushin, J. Chem. Phys. **112**, 8347–8362 (2000)
14. M. Ben-Nun, T.J. Martínez, J. Am. Chem. Soc. **122**, 6299–6300 (2000)
15. A. Hofmann, R. de Vivie-Riedle, J. Chem. Phys. **112**, 5054–5059 (2000)
16. M. Garavelli, C.S. Page, P. Celani, M. Olivucci, W.E. Schmid, S.A. Trushin, W. Fuß, J. Phys. Chem. A **105**, 4458–4469 (2001)
17. R.C. Dudek, P.M. Weber, J. Phys. Chem. A **105**, 4167–4171 (2001)
18. N. Kuthirummal, F.M. Rudakov, C.L. Evans, P.M. Weber, J. Chem. Phys. **125**, 133307 (2006)
19. F.M. Rudakov, P.M. Weber, Chem. Phys. Lett. **470**, 187–190 (2009)
20. J. Bao, M.P. Minitti, P.M. Weber, J. Phys. Chem. A **115**, 1508–1515 (2011)
21. S. Deb, P.M. Weber, Annu. Rev. Phys. Chem. **62**, 19–39 (2011)
22. O. Schalk, A.E. Boguslavskiy, A. Stolow, M.S. Schuurman, J. Am. Chem. Soc. **133**, 16451–16458 (2011)
23. S.A. Trushin, S. Diemer, W. Fuß, K.L. Kompa, W.E. Schmid, Phys. Chem. Chem. Phys. **1**, 1431–1440 (1999)
24. W. Fuß, W.E. Schmid, S.A. Trushin, J. Am. Chem. Soc. **123**, 7101–7108 (2001)
25. W. Fuß, W.E. Schmid, S.A. Trushin, Chem. Phys. **316**, 225–234 (2005)
26. R.B. Woodward, R. Hoffmann, *The Conservation of Orbital Symmetry* (Verlag Chemie, Weinheim, 1970)
27. W.L. Dilling, Chem. Rev. **69**, 845–877 (1969)
28. L. Salem, Science **191**, 822–830 (1976)
29. B.S. Hudson, B.E. Kohler, K. Schulten, Linear polyene electronic structure and potential surfaces, in *Excited States*, vol. 6, ed. by E.C. Lim (Academic Press, New York, 1982), pp. 1–95
30. V. Vaida, R.E. Turner, J.L. Casey, S.D. Colson, Chem. Phys. Lett. **54**, 25–29 (1978)
31. L.J. Rothberg, D.P. Gerrity, V. Vaida, J. Chem. Phys. **73**, 5508–5513 (1980)
32. J.P. Doering, R. McDiarmid, J. Chem. Phys. **73**, 3617–3624 (1980)
33. J.P. Doering, R. McDiarmid, J. Chem. Phys. **75**, 2477–2478 (1981)
34. R. McDiarmid, J.P. Doering, Chem. Phys. Lett. **88**, 602–606 (1982)
35. R.R. Chadwick, D.P. Gerrity, B.S. Hudson, Chem. Phys. Lett. **115**, 24–28 (1985)
36. R.R. Chadwick, M.Z. Zgierski, B.S. Hudson, J. Chem. Phys. **95**, 7204–7211 (1991)
37. B.G. Levine, T.J. Martínez, J. Phys. Chem. A **113**, 12815–12824 (2009)
38. L.W. Pickett, E. Paddock, E. Sackter, J. Am. Chem. Soc. **63**, 1073–1077 (1941)
39. R.P. Frueholz, W.M. Flicker, O.A. Mosher, A. Kuppermann, J. Chem. Phys. **70**, 2003–2013 (1979)
40. A. Sabljić, R. McDiarmid, J. Chem. Phys. **93**, 3850–3855 (1990)
41. R. McDiarmid, A. Gedanken, J. Chem. Phys. **95**, 2220–2221 (1991)
42. R. McDiarmid, A. Sabljić, J.P. Doering, J. Chem. Phys. **83**, 2147–2152 (1985)
43. Q.-Y. Shang, B.S. Hudson, Chem. Phys. Lett. **183**, 63–68 (1991)
44. F.M. Rudakov, P.M. Weber, J. Phys. Chem. A **114**, 4501–4506 (2010)
45. O. Schalk, A.E. Boguslavskiy, A. Stolow, J. Phys. Chem. A **114**, 4058–4064 (2010)
46. R.S. Mulliken, Rev. Mod. Phys. **14**, 265–274 (1942)
47. M.B. Robin, *Higher Excited States of Polyatomic Molecules*, vol. 2 (Academic Press, New York, NY, 1975)
48. P.C. Hariharan, J.A. Pople, Theor. Chim. Acta **28**, 213–222 (1973)

49. M.M. Francl, W.J. Pietro, W.J. Hehre, J.S. Binkley, M.S. Gordon, D.J. DeFrees, J.A. Pople, J. Chem. Phys. **77**, 3654–3665 (1982)
50. Y.J. Bomble, K.W. Sattelmeyer, J.F. Stanton, J. Gauss, J. Chem. Phys. **121**, 5236–5240 (2004)
51. H. Nakatsuji, O. Kitao, T. Yonezawa, J. Chem. Phys. **83**, 723–734 (1985)
52. L. Serrano-Andrés, M. Merchán, I. Nebot-Gil, B.O. Roos, M. Fülscher, J. Am. Chem. Soc. **115**, 6184–6197 (1993)
53. H. Nakano, T. Tsuneda, T. Hashimoto, K. Hirao, J. Chem. Phys. **104**, 2312–2320 (1996)
54. J.D. Watts, S.R. Gwaltney, R.J. Bartlett, J. Chem. Phys. **105**, 6979–6988 (1996)
55. M. Schreiber, M.R. Silva-Junior, S.P.A. Sauer, W. Thiel, J. Chem. Phys. **128**, 134110 (2008)
56. J. Shen, S. Li, J. Chem. Phys. **131**, 174101 (2009)
57. M.J. Davis, E.J. Heller, J. Chem. Phys. **80**, 5036–5048 (1984)
58. K.B. Møller, A.H. Zewail, Chem. Phys. Lett. **295**, 1–10 (1998)
59. K.B. Møller, N.E. Henriksen, A.H. Zewail, J. Chem. Phys. **113**, 10477–10485 (2000)

Chapter 9
Dithiane

Parts of the results described in text passages and/or figures and tables in this chapter are adapted with permission from [7]. Copyright 2012 American Chemical Society.

The disulfide bond seemingly represents a case where photochemistry efficiently competes with energy dissipation by internal conversion. It is well-recognized that upon the absorption of light a disulfide bond will cleave leaving two free radicals [1]. This process has been observed to occur on a sub ps timescale [2]. Despite this photolability, the disulfide bond formed by the oxidation of two cysteine amino acids is an important factor in determining the tertiary structure of proteins along with hydrogen bonding and hydrophobic interactions [3]. For the disulfide bond to prevail, the radicals formed by the bond cleavage must be confined such as to allow for recombination and thereby reformation of the disulfide bond. In solution, the solvent shell surrounding radicals formed by the photocleavage of a disulfide bond has been argued to provide such confinement, but a quantum yield of 30 % is still observed in some cases [2, 4–6]. In the case of proteins, this raises the question whether other parameters such as intrinsic cyclic structural motifs in the tertiary structure could play a decisive role in funneling the disulfide back to the ground state by internal conversion. We have inhere focused on the cyclic molecule dithiane as a model compound for such a structural motif, and we will focus on the dynamics unfolding subsequent to excitation to the S_1 state.

9.1 Electronic Structure

The lowest two singlet excited states of dithiane derive from promotion of an electron from a lonepair at either one of the two sulphur atoms to an antibonding orbital of the S–S bond. Due to the interaction between the two sulphur atoms, these states are not degenerate but are split at the Franck-Condon geometry by 1.0 eV according to our SA-3-CAS(10,8)SCF/6-31G(d,p) calculations. These calculations employ an active

T. S. Kuhlman, *The Non-Ergodic Nature of Internal Conversion*, Springer Theses, DOI: 10.1007/978-3-319-00386-3_9, © Springer International Publishing Switzerland 2013

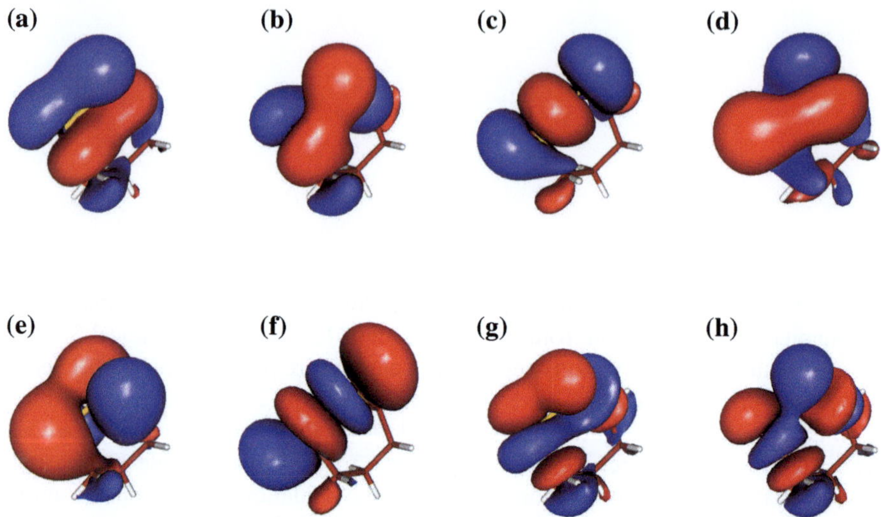

Scheme 9.1 Photolytic cleavage of the S–S bond in dithiane. Adapted with permission from [7]. Copyright 2012 American Chemical Society

(a)   **(b)**   **(c)**   **(d)**

(e)   **(f)**   **(g)**   **(h)**

Fig. 9.1 The eight orbitals included in the SA-3-CAS(10,8)SCF calculations in order of increasing energy. (**a**) σ_{SC}. (**b**) σ_{SC}. (**c**) σ_{SS}. (**d**) n. (**e**) n. (**f**) σ_{SS}^*. (**g**) σ_{SC}^*. (**h**) σ_{SC}^*

space of the eight orbitals shown in Fig. 9.1 which includes σ^* orbitals to describe possible S–C and S–S bond breakage (Scheme 9.1).

The linearly interpolated S_1 potential energy surface connecting the Franck-Condon geometry with the S_1 minimum reveals no barriers on the very steep surface. At the S_1 minimum, the S–S bond is extended by 1.5–3.6 Å, cf. Fig. 9.2. Thus, without any opposing force due to confinement, the bond would readily dissociate forming the diradical species. As the S–S bond elongates, the interaction between the two sulphur atoms decreases, and the S_1 and S_2 states become almost degenerate. Near the minimum of the S_1 potential energy surface, we have located a minimum energy conical intersection (MECI) connecting S_1 with the ground state. The geometries of the S_1 minimum and the $S_1 S_0$ MECI are closely related as only a slight torsion of the $-CH_2S^\bullet$ moieties can take one into the other, cf. Fig. 9.3. This proximity of the two points on the S_1 potential energy surface gives rise to the sloped geometry of the MECI illustrated in Fig. 9.4. Thus, once the molecule starts moving towards the S_1 minimum, the region near the conical intersection will also be visited. If nuclear motion restricts the molecule to the vicinity of this region of large coupling, the molecule should be able to return to the ground state and reform the S–S bond avoiding diradical formation. Such nuclear motion is exactly revealed by the data obtained from time-resolved mass spectrometry (TRMS).

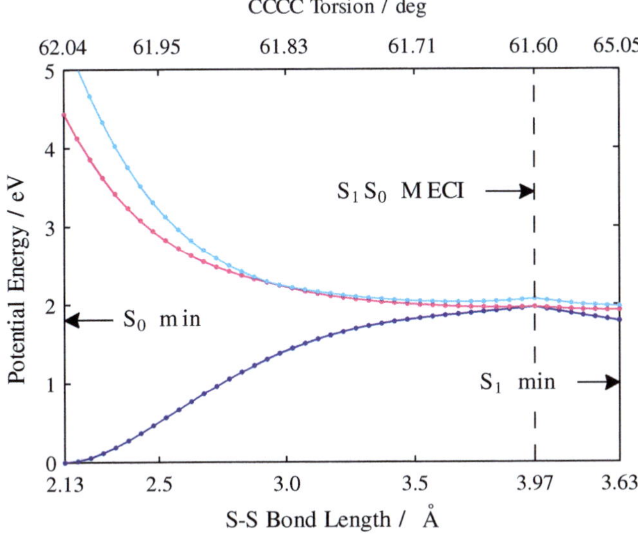

Fig. 9.2 Adiabatic potential energy surfaces for the three lowest singlet states of dithiane. The structures have been linearly interpolated in internal coordinates from the S_0 minimum to the $S_1 S_0$ MECI and from the latter to the S_1 minimum. Adapted with permission from [7]. Copyright 2012 American Chemical Society

Fig. 9.3 Geometries of dithiane with important structural parameters indicated. (**a**) The normal mode displacement vector for the lowest frequency vibrational mode is indicated by arrows on the lower structure. Adapted with permission from [7]. Copyright 2012 American Chemical Society. (**a**) S_1 minimum. (**b**) $S_1 S_0$ conical intersection

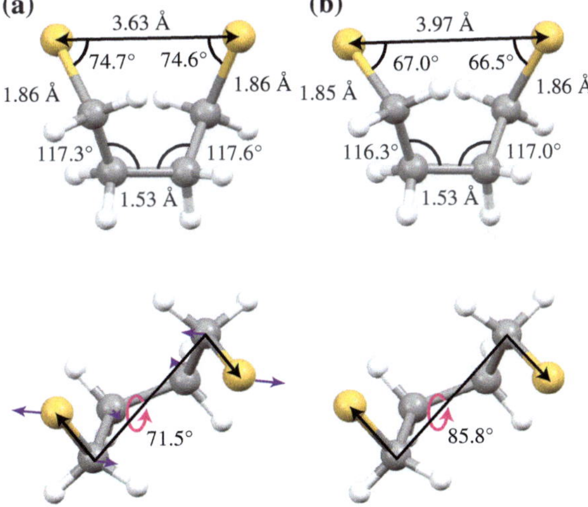

9.2 Time-Resolved Mass Spectrometry

The time in the vicinity of the MECI necessary for efficient transfer back to the ground state is provided by the cyclic structure of the molecule. The timescale for activating modes that lead to the unfolding of the carbon chain and thereby diradical

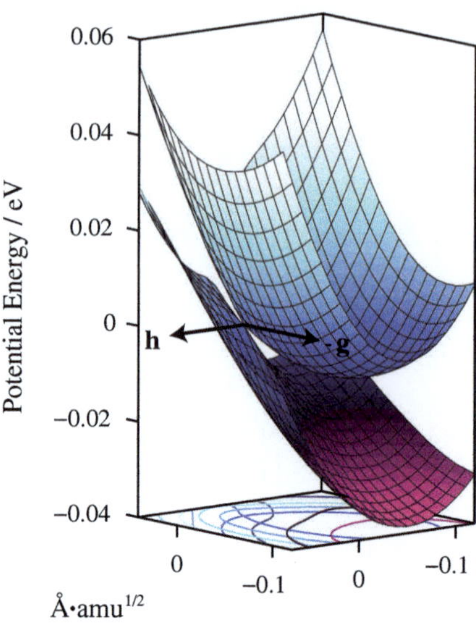

Fig. 9.4 MECI between the S_1 and S_0 states in the branching space of the scaled derivative coupling and gradient difference vectors. The intersection is observed to be sloped with an adjacent local minimum on the S_1 surface

formation follows that of internal vibrational energy redistribution. This timescale is generally on the order of several ps and longer giving the molecule amble time to return to the ground state and reform the S–S bond. Upon excitation to the S_1 state, the elongation of the S–S bond is accompanied by torsion in the carbon backbone which leads to a scissoring motion of the sulphur atoms. This motion is directly observed in the temporal evolution of the ion current of the $m/z = 55$ fragment of dithiane presented in Fig. 9.5. In the TRMS experiments, dithiane was excited to the S_1 state by a 110 fs pulse at 284 nm and probed at a later time by a 400 nm pulse of similar time duration.[1] When the sulphur atoms are in close proximity, the ionization of dithiane is enhanced giving rise to a peak in the ion current as a positive charge on one of the sulphur atoms can be stabilized by the lone pair on the other. When the sulphur atoms are far a part, this stabilization is absent, and ionization is suppressed given rise to a trough in the ion current.

The fit to the temporal evolution of the ion current reveals that the S_1 minimum is reached within <200 fs as given by the initial 177 ± 17 fs decay corresponding to the S–S bond stretch. The oscillatory component of the signal has a period of 411 ± 27 fs which is nearly identical to the period of the lowest vibrational mode in the S_1 minimum as calculated by a harmonic normal mode analysis to be 416 fs. This mode involves torsion in the carbon backbone which results in the $-CH_2S^\bullet$ moieties moving relative to one another as indicated in Fig. 9.3a. This is exactly the motion that connects the S_1 minimum with the MECI, and it can thereby induce the transition back to the ground state. This transition is also reflected in the experimental signal by the second decay component with a fitted lifetime of 2.75 ± 0.23 ps.

---

[1] The author did not take part in conducting the experiments.

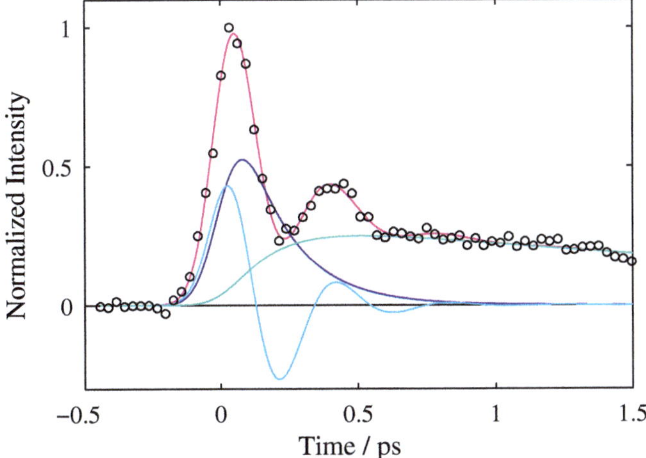

Fig. 9.5 Temporal evolution of the ion current for the $m/z = 55$ fragment of dithiane (o) along with the modulated sequential biexponential fit (pink). Also shown is the individual components of the fit: first exponential decay (purple), second exponential decay (green), and oscillatory component (blue). Adapted with permission from [7]. Copyright 2012 American Chemical Society

9.3 Conclusion

The reformation of the S–S bond in excited dithiane is the result of a non-ergodic process. Only two degrees of freedom actively take part in the process: the S–S bond stretch and torsion in the carbon backbone giving rise to a scissoring motion of the $-CH_2S^\bullet$ moieties. The bond reformation prevails as motion in these two degrees of freedom actively takes the molecule to a region of large interaction between S_1 and S_0. Hereby, the internal conversion process can effectively compete with internal vibrational energy redistribution to degrees of freedom which could lead to unfolding of the ring and ultimately result in diradical formation. The cyclic structural motif of dithiane ultimately protects the molecule from bond breakage, and a similar effect is expected to be at play in proteins.

References

1. W.E. Lyons, Nature **162**, 1004 (1948)
2. N.P. Ernsting, Chem. Phys. Lett. **166**, 221–226 (1990)
3. T.E. Creighton, BioEssays **8**, 57–63 (1988)
4. T. Bultmann, N.P. Ernsting, J. Phys. Chem. **100**, 19417–19424 (1996)
5. Y. Hirata, Y. Niga, S. Makita, T. Okada, J. Phys. Chem. A **101**, 561–565 (1997)
6. L. Milanesi, G.D. Reid, G.S. Beddard, C.A. Hunter, J.P. Waltho, Chem. Eur. J. **10**, 1705–1710 (2004)
7. A. Stephansen, R.Y. Brogaard, T.S. Kuhlman, L.B. Klein, J.B. Christensen, T.I. Sølling, J. Am. Chem. Soc. **134**, 20279–20281 (2012)

Part V
Concluding Remarks

Chapter 10
Summarizing Discussion

The work presented in this thesis exhibit the complexity of the interplay between nuclear dynamics and electronic transitions. The main lesson is that internal conversion in small polyatomic molecules is truly a dynamical process. Specific nuclear motion is instrumental in directing the system towards configurations where coupling to other electronic states is large and wherefrom population transfer can proceed. This in turn encompasses the concept of the dynamics being non-ergodic or non-statistical—the observation of coherent dynamics entails that the timescale of internal conversion is shorter than that of internal vibrational energy redistribution. This direct observation of non-ergodic behavior is an immediate consequence of the preparation of the system by a femtosecond laser pulse resulting in an initial localization in space and energy. It should be emphasized that the non-ergodicity is a function of the molecule and not of the activation process—the excited state dynamics of the molecules under investigation are inherently non-ergodic irrespective of the preparation process. We observed different effects of the non-ergodic behavior in the different molecular systems under investigation.

From the experimental study on the cycloketones, we observed that the timescale for the $(n, 3s) \rightarrow (n, \pi^*)$ internal conversion ranges over more than an order of magnitude from 0.37 ps in 2-methylcyclobutanone to 9.67 ps in cyclohexanone. It was revealed from the experimental data that the nuclear dynamics leading to this internal conversion are localized in real space in proximity of the carbonyl chromophore. Alkyl substitution and change of the size of the ring, which locally affects the structure in vicinity of the carbonyl chromophore, both have strong effects on the rate due to this locality. Effectively, these structural changes affect the magnitude of the coupling between the electronic states by determining the region of configuration space available to the molecule on the $(n, 3s)$ state. A larger available configuration space allows the molecule to visit regions of larger coupling resulting in a faster rate of internal conversion. The effects of the structural differences can be quantified by primarily two parameters: the frequency of the central ring-puckering vibration and the energy difference between the Franck-Condon and equilibrium geometries in the $(n, 3s)$ state. A lower frequency and a larger energy difference result in a faster rate

T. S. Kuhlman, *The Non-Ergodic Nature of Internal Conversion*, Springer Theses, DOI: 10.1007/978-3-319-00386-3_10, © Springer International Publishing Switzerland 2013

of transition. On the other hand, an increase in the size of phase space and the density of vibrational states as obtained by alkyl substitution away from the carbonyl group does not significantly affect the rate of internal conversion. This cements the non-ergodicity of the internal conversion process revealed by the observation of coherent vibrational motion—the molecule does not explore the full phase space in the excited state before population transfer proceeds. Had this been the case, these substitutions would have had a much larger effect on the rate of transition.

The importance of low-frequency modes was to some extent confirmed by the wavepacket simulations on cyclobutanone and cyclopentanone. More importantly, the simulations revealed the importance of point group symmetry. For cyclobutanone, internal conversion is a direct process where motion in Franck-Condon active coordinates mediates a prompt population transfer. In cyclopentanone, the point group symmetry dictates that internal conversion is an indirect process where internal vibrational energy redistribution is a bottleneck for activating coupling modes. Consequently, the component of prompt population transfer in cyclopentanone is very small compared to the component of delayed population transfer. However, as the experimental timescale for population transfer in cyclopentanone is not reproduced by the simulations, it is very conceivable that the five mode model is not sufficient in this case to fully describe the internal conversion process.

Similar to the case of the cycloketones, coherent dynamics were also observed in the cyclopentadienes on the timescale of the internal conversion process which ranges by a factor of approximately 3 from ~ 50 fs in cyclopentadiene to ~ 150 fs in hexamethylcyclopentadiene. Whereas the timescale differences in the cycloketones in essence is a consequence of a difference in the magnitude of the coupling between the electronic states, the timescale differences for the cyclopentadienes result from differences in the timescale for reaching the configuration where the coupling is large. As also observed for the cycloketones, the nuclear dynamics leading to internal conversion in the cyclopentadienes are localized in real space in this case in the bond alternation and double bond twist coordinates. The latter motion takes the molecules from a planar configuration, where the coupling to the ground state is very low, to the vicinity of the conical intersection seam, where the coupling is very large. The calculations do not include the extra degrees of freedom of the methyl groups in Me_4CPD and Me_6CPD and the differing timescales are truly a result of a kinematic effect and not a result of a change in the size of phase space. Once again, phase space is open to the molecule but only a small region is accessed leading to the internal conversion exhibiting a non-ergodic nature.

The ultrafast internal conversion in the cyclopentadienes results from the interaction between the dark doubly excited state and the bright singly excited state. As the molecule leaves the Franck-Condon region on the steep potential energy surface along out-of-plane modes, it enters a region on the surface where mixing of singly and doubly excited character in the two lowest excited valence states occurs. The nuclear motion is reflected by a decaying band in the time-resolved photoelectron spectra which resembles the appearance of a band decaying due to a non-adiabatic transition. However, the decay is indeed a consequence of the closing of the probe window due to the rapidly increasing ionization potential. The full extent of the coherent dynamics are revealed in the spectrum if a higher frequency probe is employed as the wavepacket can then be followed on the entire excited-state potential energy surface.

In dithiane, the reduced-space dynamics are reflected in the conservation of the disulfide bond following light absorption. The significant stretching of the disulfide bond upon excitation leads the molecule straight towards the minimum of the excited-state potential energy surface in the region of the conical intersection. Whereas the stretching motion initially induces the coupling to the ground state, it is the accompanying coherent backbone torsion which ensures recurrent motion between the minimum and the conical intersection allowing amble time for population transfer to proceed. Had a larger region of phase space been sampled during the process, the unfolding of the carbon chain would have led to the formation of the diradical. In addition, had the coupling between the excited and ground states been significantly smaller, a prolonged lifetime in the excited state could have allowed for internal vibrational energy redistribution to proceed. Thereby, the vibrational modes which lead to unfolding of the chain would have been activated and diradical formation would again prevail. That this is not observed is a testament to the non-ergodic nature of the process.

In the introduction, we reflected on the possibility of identifying the main parameters that determine the timescale for a given internal conversion process. In the Born-Oppenheimer approximation, it is ultimately the electronic structure that determines the potential energy surfaces on which the nuclear dynamics take place, and one could therefore argue that this is the main parameter. On the other hand, the non-adiabatic couplings leading to the breakdown of the approximation are due to the nuclear kinetic energy operator highlighting the central role of nuclear dynamics in their own right. From one perspective, the electronic structure defines the potential energy surfaces, but it is the nuclear dynamics that determine which parts of these surfaces are visited and thus of importance for electronic state-transfer processes. A common trait of the three types of molecules investigated in our work is the involvement of very few degrees of freedom in the nuclear dynamics that lead to internal conversion. The activation and frequency of as well as energy release into these ultimately determine on what timescale and to what extent the region of the potential energy surface with large coupling to other states is visited and thereby the timescale for internal conversion. This connection is evidenced by the significant effect of selective modification of these degrees of freedom on the rate of internal conversion observed in our work. The localization of the dynamics furthermore allows for a connection between nuclear dynamics and structural elements. In the cycloketones, the main structural element of importance is the C-CO-C moiety, in the cyclopentadienes it is the double bonds, and in dithiane it is the disulfide bond. In essence, simply from a consideration of molecular structure, the relative rates of internal conversion in related molecules can to some extent be predicted. We hope that these concepts will provide a useful basis on which to undertake and discuss future work.

Appendix A
Supporting Information for Chapter 7

A.1 Diffuse Basis Functions

The exponents of the diffuse functions are given in Table A.1 [1]. Tables A.2 and A.3 collect the contraction coefficients for three sets of diffuse functions for cyclobutanone and cyclopentanone determined following the prescription in Ref. [2].

Table A.1 Exponents of the primitive basis functions for three values of angular momentum l and eight values of the principal quantum number n

n	$l = 0\ (s)$	$l = 1\ (p)$	$l = 2\ (d)$
2.0	0.02462393	0.04233528	0.06054020
2.5	0.01125334	0.01925421	0.02744569
3.0	0.00585838	0.00998821	0.01420440
3.5	0.00334597	0.00568936	0.00807659
4.0	0.00204842	0.00347568	0.00492719
4.5	0.00132364	0.00224206	0.00317481
5.0	0.00089310	0.00151064	0.00213712
5.5	0.00062431	0.00105475	0.00149102

T. S. Kuhlman, *The Non-Ergodic Nature of Internal Conversion*, Springer Theses, DOI: 10.1007/978-3-319-00386-3, © Springer International Publishing Switzerland 2013

Table A.2 Contraction coefficients for three sets of diffuse functions for cyclobutanone

n	$l = 0$ (s)		
2.0	0.18757690	−0.66901176	0.84247489
2.5	1.02138077	−0.78473105	0.22173456
3.0	−0.43425367	1.16965432	−2.06764134
3.5	0.36093788	0.54692212	−0.48159425
4.0	−0.02660994	−0.06342005	1.66260101
4.5	−0.28561643	0.02697933	0.84767931
5.0	0.29491969	0.02931525	−0.64237097
5.5	−0.10167962	−0.01953191	0.24267274
n	$l = 1$ (p)		
2.0	0.28733454	−0.51482697	0.51694331
2.5	0.59336328	−0.44690850	0.16166798
3.0	0.01309852	0.66233366	−1.02008576
3.5	0.43319267	−0.28557390	0.52617583
4.0	−0.52217482	1.71767160	−2.19954435
4.5	0.49010121	−1.54212007	3.90540389
5.0	−0.28532868	0.97402587	−1.77209367
5.5	0.07604473	−0.26767560	0.57951653
n	$l = 2$ (d)		
2.0	0.13417940	−0.21298673	0.23409036
2.5	0.26292179	−0.24606889	0.19986112
3.0	0.36090055	−0.23334084	0.05457146
3.5	0.32203482	−0.06405318	0.00085129
4.0	0.07999795	0.47688100	−0.96575675
4.5	−0.00353851	0.50476530	0.07905963
5.0	0.02622863	0.02637718	0.11480124
5.5	−0.01171333	0.07072576	1.02575722

Table A.3 Contraction coefficients for three sets of diffuse functions for cyclopentanone

n	$l = 0$ (s)		
2.0	0.00418467	−0.38848697	0.55908927
2.5	1.18993678	−1.26254802	0.90072528
3.0	−0.46431864	1.24271128	−2.21416585
3.5	0.41221310	0.59751973	−0.98671842
4.0	−0.02661507	−0.01247047	1.70203509
4.5	−0.33434677	0.09202366	1.06594946
5.0	0.34380467	−0.04535688	−0.64214518
5.5	−0.11840101	0.00656561	0.24234048
n	$l = 1$ (p)		
2.0	0.27725230	−0.61352243	0.65451483
2.5	0.53090715	−0.27833537	−0.10891805
3.0	0.12051661	0.42258781	−0.73894617
3.5	0.35196052	0.00376434	0.12462769
4.0	−0.41598528	1.34057797	−1.44956674
4.5	0.39835946	−1.18216962	3.11862406
5.0	−0.23366103	0.75511770	−1.31887282
5.5	0.06236085	−0.20643878	0.43609035
n	$l = 2$ (d)		
2.0	0.16439619	−0.32105070	0.35040189
2.5	0.27597911	−0.27560551	0.19803434
3.0	0.33584271	−0.14156831	−0.09950192
3.5	0.31715460	0.01710969	−0.16583083
4.0	0.08042406	0.47698256	−0.78001504
4.5	−0.00479743	0.43481886	0.20066895
5.0	0.02559397	−0.01114904	0.44671880
5.5	−0.01121493	0.06198461	0.62692971

A.2 Equilibrium Geometries

The ground, (n, π^*), and $(n, 3s)$ equilibrium geometries of cyclobutanone and cyclopentanone are given in Tables A.4, A.5, A.6, A.7, A.8, A.9, A.10, A.11. The ground state geometries were obtained by MP2 and CCSD with the cc-pVTZ basis set in Gaussian 03 and Gaussian 09 [3, 4]. The excited state geometries were obtained by EOM-CCSD using the extended cc-pVTZ+1s1p1d basis set in Cfour [5]. In the tables, X indicates the position of the ghost atom onto which the diffuse functions were placed. The position of the ghost atom was initially taken as the charge centroid of the lowest cationic state determined from atom-centered LoProp charges [6] calculated in Molcas [7]. Due to the very extended nature of these basis functions, the exact placement is not critical. Another choice is to place the functions on the carbon atom of the carbonyl group which was employed when calculating the harmonic frequencies at the equilibrium geometries of the excited states.

Table A.4 Ground state equilibrium geometry of cyclobutanone (in Å) obtained at the MP2/cc-pVTZ level of theory

Atom	x	y	z
C	−0.239133	−1.425773	0.000000
C	−0.239133	−0.334089	1.095225
C	0.150868	0.650486	0.000000
C	−0.239133	−0.334089	−1.095225
H	0.686707	−1.987113	0.000000
H	−1.076181	−2.109987	0.000000
H	0.434302	−0.426955	1.937860
H	−1.241142	−0.114128	1.452686
H	0.434302	−0.426955	−1.937860
H	−1.241142	−0.114128	−1.452686
O	0.675294	1.730007	0.000000

Table A.5 Ground state equilibrium geometry of cyclobutanone (in Å) obtained at the CCSD/cc-pVTZ level of theory

Atom	x	y	z
C	0.143957	−1.456543	0.000000
C	−0.091246	−0.386572	1.104995
C	−0.034336	0.674327	0.000000
C	−0.091246	−0.386572	−1.104995
H	1.164767	−1.831344	0.000000
H	−0.544168	−2.296732	0.000000
H	0.648899	−0.291705	1.896418
H	−1.086189	−0.436035	1.548271
H	0.648899	−0.291705	−1.896418
H	−1.086189	−0.436035	−1.548271
O	0.086389	1.864433	0.000000

Table A.6 (n, π^*) state equilibrium geometry of cyclobutanone (in Å) obtained at the EOM-CCSD/cc-pVTZ+1s1p1d level of theory

Atom	x	y	z
C	−1.496048	−0.012062	0.009687
C	−0.401706	−1.099674	0.102231
C	0.640057	−0.009234	−0.259848
C	−0.404282	1.077982	0.104441
H	−1.993152	−0.011982	−0.956738
H	−2.240936	−0.013703	0.799987
H	−0.444319	−1.945041	−0.580421
H	−0.235500	−1.449477	1.125215
H	−0.448620	1.924920	−0.576164
H	−0.238979	1.425844	1.128234
O	1.868130	−0.008113	0.099919
X	−0.042915	0.007156	−0.020238

Table A.7 $(n, 3s)$ state equilibrium geometry of cyclobutanone (in Å) obtained at the EOM-CCSD/cc-pVTZ+1s1p1d level of theory

Atom	x	y	z
C	−1.575325	−0.015080	−0.105112
C	−0.492191	−1.096422	−0.094997
C	0.690366	−0.014060	−0.098159
C	−0.493084	1.067216	−0.105908
H	−2.181066	−0.019842	−1.003957
H	−2.188393	−0.010861	0.788803
H	−0.347224	−1.668302	−1.015431
H	−0.352799	−1.654193	0.835027
H	−0.348704	1.629701	−1.032218
H	−0.354196	1.634555	0.818391
O	1.850375	−0.013500	−0.096769
X	−0.013362	0.010057	0.070121

Table A.8 Ground state equilibrium geometry of cyclopentanone (in Å) obtained at the MP2/cc-pVTZ level of theory

Atom	x	y	z
C	0.000000	0.764737	−1.362297
C	0.000000	−0.764737	−1.362297
C	0.524110	−1.111679	0.022999
C	0.000000	0.000000	0.911689
C	−0.524110	1.111679	0.022999
H	1.021254	1.123931	−1.469389
H	−0.587587	1.190079	−2.167216
H	0.587587	−1.190079	−2.167216
H	−1.021254	−1.123931	−1.469389
H	0.242093	−2.084007	0.407610
H	1.611593	−1.047251	0.047856
H	−1.611593	1.047251	0.047856
H	−0.242093	2.084007	0.407610
O	0.000000	0.000000	2.120465

Table A.9 Ground state equilibrium geometry of cyclopentanone (in Å) obtained at the CCSD/cc-pVTZ level of theory

Atom	x	y	z
C	−0.011438	0.769497	−1.369003
C	0.011438	−0.769497	−1.369003
C	0.524246	−1.122169	0.030365
C	0.000000	0.000000	0.922887
C	−0.524246	1.122169	0.030365
H	1.004602	1.149409	−1.498593
H	−0.623485	1.183657	−2.167885
H	0.623485	−1.183657	−2.167885

(continued)

Table A.9 (continued)

Atom	x	y	z
H	−1.004602	−1.149409	−1.498593
H	0.219765	−2.095056	0.409259
H	1.616322	−1.078547	0.066445
H	−1.616322	1.078547	0.066445
H	−0.219765	2.095056	0.409259
O	0.000000	0.000000	2.127219

Table A.10 (n, π^*) state equilibrium geometry of cyclopentanone (in Å) obtained at the EOM-CCSD/cc-pVTZ+1s1p1d level of theory

Atom	x	y	z
C	−0.800227	−0.021668	0.001795
C	0.104574	−1.280585	0.128683
C	1.491776	−0.744107	−0.231496
C	1.489495	0.709750	0.232505
C	0.099336	1.241072	−0.124127
H	−0.307791	−2.080639	−0.486044
H	−0.002155	−1.524864	1.192974
H	2.248377	−1.349248	0.270790
H	1.646790	−0.811074	−1.306170
H	2.242618	1.317694	−0.271574
H	1.646839	0.777162	1.306818
H	−0.010451	1.487949	−1.187468
H	−0.315304	2.037658	0.493653
O	−1.982671	−0.024376	0.000544
X	−0.041116	0.016828	−0.001101

Table A.11 $(n, 3s)$ state equilibrium geometry of cyclopentanone (in Å) obtained at the EOM-CCSD/cc-pVTZ+1s1p1d level of theory

Atom	x	y	z
C	−0.757421	−0.060563	−0.009309
C	0.103972	−1.267442	0.327197
C	1.504280	−0.745617	−0.014206
C	1.473168	0.710578	0.461005
C	0.062690	1.224607	0.109635
H	−0.191038	−2.154263	−0.231781
H	0.016117	−1.497068	1.398130
H	2.288573	−1.330190	0.465368
H	1.658561	−0.790574	−1.093230
H	2.257397	1.321469	0.015196
H	1.605373	0.738898	1.543802
H	0.037973	1.772128	−0.834340
H	−0.350965	1.880399	0.879873
O	−2.013939	−0.046432	0.330740
X	−0.038068	0.024636	−0.179418

A.3 Parameters of the Vibronic Coupling Hamiltonian

The parameters of the vibronic coupling Hamiltonian obtained from fitting to potential energy surfaces calculated at the EOM-CCSD/cc-pVTZ+1s1p1d level of theory are given in Tables A.12, A.13, A.14, A.15, A.16, A.17, A.18, A.19, A.20, A.21 for cyclobutanone and Tables A.22, A.23, A.24, A.25, A.26, A.27, A.28, A.29, A.30, A.31 for cyclopentanone. The electronic labels $v, w = 1, 2, 3, 4, 5$ correspond to the ground, (n, π^*), $(n, 3s)$, $(n, 3p)$, and ground cationic states respectively. The labels for the nuclear degrees of freedom $\kappa, \kappa' = 1, 2, 7, 12, 21$ for cyclobutanone and $1, 3, 8, 16, 28$ for cyclopentanone. These labels correspond to the ring-puckering, C=O out-of-plane bend (carbonyl pyramidalization), symmetric C–CO–C stretch, asymmetric C–CO–C stretch, and C=O stretch respectively.

A.3.1 Cyclobutanone

Table A.12 Vibrational frequencies ω_κ (in eV) for the normal modes of cyclobutanone

κ	1	2	7	12	21
ω_κ	0.0141	0.0501	0.1102	0.1373	0.2300

Table A.13 On-diagonal constants $E^{(v)}$ (in eV) for the five states of cyclobutanone

v	1	2	3	4	5
$E^{(v)}$	0.0000	4.4654	6.5970	7.1847	9.5254

Table A.14 Parameters for the Morse potential of the C=O stretch mode for the five states of cyclobutanone

v	1	2	3	4	5
$D_{21}^{(v)}$/eV	28.1696	19.4099	7.0043	4.5790	17.7349
$\alpha_{21}^{(v)}$	−0.0651	−0.0657	−0.1029	−0.1167	−0.0753
$Q_{21,0}^{(v)}$	−0.1140	−1.6670	0.3191	−0.0009	0.2736
$E_0^{(v)}$/eV	−0.0016	−0.2598	−0.0073	0.0000	−0.0074

Table A.15 On-diagonal linear coupling constants $\beta_\kappa^{(v)}$ (in eV) for the normal modes of cyclobutanone

κ	$\beta_\kappa^{(1)}$	$\beta_\kappa^{(2)}$	$\beta_\kappa^{(3)}$	$\beta_\kappa^{(4)}$	$\beta_\kappa^{(5)}$
1	0.0052	0.0653	0.0169	−0.0179	0.0075
2	−0.0027	−0.0074	−0.0453	−0.0096	−0.0016
7	−0.0090	−0.0024	0.0189	0.0092	0.0106
12	$\cdots$	$\cdots$	$\cdots$	$\cdots$	$\cdots$
21	$\cdots$	$\cdots$	$\cdots$	$\cdots$	$\cdots$

Table A.16 On-diagonal bilinear coupling constants $\gamma_{\kappa\kappa'}^{(\nu)}$ (in eV) for the normal modes of cyclobutanone

	1	2	7	12	21
$\gamma_{\kappa\kappa'}^{(1)}$					
1	0.0744	−0.0094	−0.0057	$\cdots$	−0.0471
2	−0.0094	0.0451	−0.0169	$\cdots$	−0.0493
7	−0.0057	−0.0169	0.0092	$\cdots$	0.0134
12	$\cdots$	$\cdots$	$\cdots$	0.0076	$\cdots$
21	−0.0470	−0.0493	0.0134	$\cdots$	$\cdots$
$\gamma_{\kappa\kappa'}^{(2)}$					
1	0.0948	−0.0336	−0.0124	$\cdots$	−0.0639
2	−0.0336	0.0115	−0.0203	$\cdots$	−0.0532
7	−0.0124	−0.0203	0.0070	$\cdots$	0.0264
12	$\cdots$	$\cdots$	$\cdots$	−0.0462	$\cdots$
21	−0.0639	−0.0532	0.0264	$\cdots$	$\cdots$
$\gamma_{\kappa\kappa'}^{(3)}$					
1	0.0900	−0.0391	−0.0161	$\cdots$	−0.0308
2	−0.0391	0.0478	−0.0192	$\cdots$	−0.0323
7	−0.0161	−0.0192	0.0084	$\cdots$	0.0369
12	$\cdots$	$\cdots$	$\cdots$	−0.0462	$\cdots$
21	−0.0308	−0.0323	0.0369	$\cdots$	$\cdots$
$\gamma_{\kappa\kappa'}^{(4)}$					
1	0.0429	−0.0499	0.0081	$\cdots$	−0.0497
2	−0.0499	0.0401	−0.0265	$\cdots$	−0.0467
7	0.0081	−0.0265	−0.0264	$\cdots$	0.0253
12	$\cdots$	$\cdots$	$\cdots$	−0.0492	$\cdots$
21	−0.0497	−0.0467	0.0253	$\cdots$	$\cdots$
$\gamma_{\kappa\kappa'}^{(5)}$					
1	0.0851	−0.0159	−0.0110	$\cdots$	−0.0491
2	−0.0159	0.0525	−0.0210	$\cdots$	−0.0425
7	−0.0110	−0.0210	0.0018	$\cdots$	0.0191
12	$\cdots$	$\cdots$	$\cdots$	−0.0479	$\cdots$
21	−0.0491	−0.0425	0.0191	$\cdots$	−0.0288

Table A.17 On-diagonal linear-quadratic coupling constants $\iota^{(v)}_{\kappa\kappa'}$ (in eV) for the normal modes of cyclobutanone

$\iota^{(1)}_{\kappa\kappa'}$	1^2	2^2	7^2	12^2	21^2
1	0.0085	$\cdots$	$\cdots$	$\cdots$	$\cdots$
2	−0.0047	−0.0038	$\cdots$	0.0024	$\cdots$
7	−0.0037	−0.0028	−0.0027	−0.0069	$\cdots$
12	$\cdots$	$\cdots$	$\cdots$	$\cdots$	$\cdots$
21	−0.0192	−0.0181	−0.0016	−0.0177	$\cdots$
$\iota^{(2)}_{\kappa\kappa'}$					
1	0.0004	0.0128	0.0082	−0.0011	0.0007
2	−0.0126	0.0032	$\cdots$	0.0033	−0.0052
7	−0.0108	−0.0001	−0.0031	−0.0079	−0.0007
12	$\cdots$	$\cdots$	$\cdots$	$\cdots$	$\cdots$
21	−0.0198	−0.0140	$\cdots$	−0.0220	−0.0016
$\iota^{(3)}_{\kappa\kappa'}$					
1	0.0095	0.0010	−0.0003	0.0001	0.0123
2	−0.0062	0.0039	−0.0025	0.0020	−0.0025
7	−0.0060	$\cdots$	−0.0029	−0.0076	−0.0009
12	$\cdots$	$\cdots$	$\cdots$	$\cdots$	$\cdots$
21	−0.0240	−0.0058	0.0012	−0.0174	−0.0025
$\iota^{(4)}_{\kappa\kappa'}$					
1	0.0039	0.0003	−0.0012	0.0019	0.0008
2	−0.0016	−0.0008	$\cdots$	0.0003	−0.0116
7	0.0024	0.0004	−0.0055	−0.0039	0.0001
12	$\cdots$	$\cdots$	$\cdots$	$\cdots$	$\cdots$
21	−0.0113	0.0022	0.0021	−0.0176	−0.0025
$\iota^{(5)}_{\kappa\kappa'}$					
1	0.0075	$\cdots$	$\cdots$	0.0002	$\cdots$
2	−0.0046	−0.0051	$\cdots$	0.0019	$\cdots$
7	−0.0057	−0.0037	−0.0008	−0.0084	$\cdots$
12	$\cdots$	$\cdots$	$\cdots$	$\cdots$	$\cdots$
21	−0.0164	−0.0165	−0.0007	−0.0199	$\cdots$

Table A.18 On-diagonal quartic coupling constants $\epsilon^{(v)}_{\kappa}$ (in eV) for the normal modes of cyclobutanone

κ	$\epsilon^{(1)}_{\kappa}$	$\epsilon^{(2)}_{\kappa}$	$\epsilon^{(3)}_{\kappa}$	$\epsilon^{(4)}_{\kappa}$	$\epsilon^{(5)}_{\kappa}$
1	0.0344	0.0308	0.0219	0.0315	0.0324
2	0.0082	0.0070	0.0018	0.0034	0.0079
7	0.0003	$\cdots$	−0.0015	0.0004	0.0006
12	0.0005	0.0017	0.0017	−0.0001	0.0029
21	$\cdots$	$\cdots$	$\cdots$	$\cdots$	$\cdots$

Table A.19 Off-diagonal linear coupling constants $\lambda_\kappa^{(v,w)}$ (in eV) for the normal modes of cyclobutanone

κ	$\lambda_\kappa^{(2,3)}$	$\lambda_\kappa^{(2,4)}$	$\lambda_\kappa^{(3,4)}$
1	0.0693	0.2284	0.0024
2	0.2251	0.1194	−0.0183
7	−0.0287	−0.0678	−0.0448
12	$\cdots$	$\cdots$	$\cdots$
21	−0.0663	−0.0316	−0.0025

Table A.20 Off-diagonal bilinear coupling constants $\mu_{\kappa\kappa'}^{(v,w)}$ (in eV) for the normal modes of cyclobutanone

$\mu_{\kappa\kappa'}^{(2,3)}$	1	2	7	12	21
1	−0.0375	0.0342	0.0344	$\cdots$	0.0271
2	0.0342	−0.0043	−0.0085	$\cdots$	0.0143
7	0.0344	−0.0085	0.0025	$\cdots$	0.0148
12	$\cdots$	$\cdots$	$\cdots$	−0.0007	$\cdots$
21	0.0271	0.0143	0.0148	$\cdots$	−0.0154
$\mu_{\kappa\kappa'}^{(2,4)}$					
1	0.0116	−0.0019	−0.0090	$\cdots$	0.0009
2	−0.0019	−0.0042	0.0051	$\cdots$	0.0092
7	−0.0090	0.0051	0.0002	$\cdots$	−0.0010
12	$\cdots$	$\cdots$	$\cdots$	−0.0023	$\cdots$
21	0.0009	0.0092	−0.0010	$\cdots$	0.0038
$\mu_{\kappa\kappa'}^{(3,4)}$					
1	−0.0011	−0.0072	−0.0013	$\cdots$	−0.0056
2	−0.0072	−0.0059	0.0033	$\cdots$	0.0082
7	−0.0013	0.0033	−0.0008	$\cdots$	−0.0093
12	$\cdots$	$\cdots$	$\cdots$	0.0008	$\cdots$
21	−0.0056	0.0082	−0.0093	$\cdots$	0.0093

Table A.21 Off-diagonal linear-quadratic coupling constants $\eta_{\kappa\kappa'}^{(v,w)}$ (in eV) for the normal modes of cyclobutanone

$\eta_{\kappa\kappa'}^{(2,3)}$	1^2	2^2	7^2	12^2	21^2
1	−0.0033	⋯	⋯	−0.0005	0.0001
2	−0.0022	−0.0109	0.0021	−0.0007	⋯
7	−0.0018	−0.0025	0.0003	⋯	0.0014
12	⋯	⋯	⋯	⋯	⋯
21	0.0003	−0.0002	0.0025	0.0050	−0.0019
$\eta_{\kappa\kappa'}^{(2,4)}$					
1	−0.0113	⋯	⋯	−0.0014	⋯
2	−0.0072	−0.0081	⋯	−0.0011	⋯
7	0.0026	0.0012	0.0027	−0.0005	⋯
21	⋯	⋯	⋯	⋯	⋯
21	−0.0073	−0.0023	0.0015	0.0002	0.0003
$\eta_{\kappa\kappa'}^{(3,4)}$					
1	0.0057	⋯	⋯	−0.0007	⋯
2	−0.0034	0.0033	⋯	−0.0013	⋯
7	−0.0016	0.0009	0.0017	⋯	0.0001
12	⋯	⋯	⋯	⋯	⋯
21	−0.0044	−0.0024	−0.0016	−0.0003	0.0008

A.3.2 Cyclopentanone

Table A.22 Vibrational frequencies ω_κ (in eV) for the normal modes of cyclopentanone

κ	1	3	8	16	28
ω_κ	0.0121	0.0561	0.1038	0.1473	0.2241

Table A.23 On-diagonal constants $E^{(v)}$ (in eV) for the five states of cyclopentanone

v	1	2	3	4	5
$E^{(v)}$	0.0000	4.3500	6.4804	7.0063	9.2977

Table A.24 Parameters for the Morse potential of the C=O stretch mode for the five states of cyclopentanone

v	1	2	3	4	5
$D_{28}^{(v)}$/eV	40.8297	31.0856	6.7409	5.1947	18.3302
$\alpha_{28}^{(v)}$	-0.0478	-0.0468	-0.0988	-0.1038	-0.0673
$Q_{28,0}^{(v)}$	-0.1677	-1.9098	0.3899	0.4137	0.3126
$E_0^{(v)}$/eV	-0.0026	-0.2718	-0.0096	-0.0092	-0.0079

Table A.25 On-diagonal linear coupling constants $\beta_\kappa^{(v)}$ (in eV) for the normal modes of cyclopentanone

κ	$\beta_\kappa^{(1)}$	$\beta_\kappa^{(2)}$	$\beta_\kappa^{(3)}$	$\beta_\kappa^{(4)}$	$\beta_\kappa^{(5)}$
1	$\cdots$	$\cdots$	$\cdots$	$\cdots$	$\cdots$
3	$\cdots$	$\cdots$	$\cdots$	$\cdots$	$\cdots$
8	-0.0288	-0.0445	-0.0543	-0.0543	-0.0644
16	$\cdots$	$\cdots$	$\cdots$	$\cdots$	$\cdots$
28	$\cdots$	$\cdots$	$\cdots$	$\cdots$	$\cdots$

Table A.26 On-diagonal bilinear coupling constants $\gamma_{\kappa\kappa'}^{(v)}$ (in eV) for the normal modes of cyclopentanone

$\gamma_{\kappa\kappa'}^{(1)}$	1	3	8	16	28
1	0.0350	0.0119	$\cdots$	0.0074	$\cdots$
3	0.0119	0.0183	$\cdots$	-0.0108	$\cdots$
8	$\cdots$	$\cdots$	0.0175	$\cdots$	0.0067
16	0.0074	-0.0109	$\cdots$	0.0103	$\cdots$
28	$\cdots$	$\cdots$	0.0067	$\cdots$	$\cdots$
$\gamma_{\kappa\kappa'}^{(2)}$					
1	0.0049	-0.0214	$\cdots$	-0.0247	$\cdots$
3	-0.0214	-0.0372	$\cdots$	0.0196	$\cdots$
8	$\cdots$	$\cdots$	0.0196	$\cdots$	0.0034
16	-0.0247	0.0196	$\cdots$	0.0242	$\cdots$
28	$\cdots$	$\cdots$	0.0034	$\cdots$	$\cdots$
$\gamma_{\kappa\kappa'}^{(3)}$					
1	0.0447	0.0109	$\cdots$	0.0185	$\cdots$
3	0.0109	0.0139	$\cdots$	-0.0107	$\cdots$
8	$\cdots$	$\cdots$	0.0046	$\cdots$	-0.0054
16	0.0185	-0.0107	$\cdots$	-0.0713	$\cdots$
28	$\cdots$	$\cdots$	-0.0054	$\cdots$	$\cdots$
$\gamma_{\kappa\kappa'}^{(4)}$					
1	0.0101	0.0135	$\cdots$	0.0189	$\cdots$
3	0.0135	0.0123	$\cdots$	-0.0053	$\cdots$
8	$\cdots$	$\cdots$	-0.0012	$\cdots$	-0.0047
16	0.0189	-0.0053	$\cdots$	-0.0358	$\cdots$

(continued)

Table A.26 (continued)

$\gamma_{\kappa\kappa'}^{(1)}$	1	3	8	16	28
28	$\cdots$	$\cdots$	-0.0047	$\cdots$	$\cdots$
$\gamma_{\kappa\kappa'}^{(5)}$					
1	0.0220	0.0194	$\cdots$	0.0077	$\cdots$
3	0.0194	0.0055	$\cdots$	0.0018	$\cdots$
8	$\cdots$	$\cdots$	0.0085	$\cdots$	0.0010
16	0.0077	0.0018	$\cdots$	-0.0323	$\cdots$
28	$\cdots$	$\cdots$	0.0010	$\cdots$	-0.0581

Table A.27 On-diagonal linear-quadratic coupling constants $\iota_{\kappa\kappa'}^{(v)}$ (in eV) for the normal modes of cyclopentanone

$\iota_{\kappa\kappa'}^{(1)}$	1^2	3^2	8^2	16^2	28^2
1	$\cdots$	$\cdots$	$\cdots$	$\cdots$	$\cdots$
3	$\cdots$	$\cdots$	$\cdots$	$\cdots$	$\cdots$
8	0.0056	0.0021	-0.0096	-0.0073	-0.0012
16	$\cdots$	$\cdots$	$\cdots$	$\cdots$	$\cdots$
28	-0.0230	-0.0048	-0.0020	-0.0079	$\cdots$
$\iota_{\kappa\kappa'}^{(2)}$					
1	$\cdots$	$\cdots$	$\cdots$	$\cdots$	$\cdots$
3	$\cdots$	$\cdots$	$\cdots$	$\cdots$	$\cdots$
8	0.0077	0.0045	-0.0082	-0.0018	$\cdots$
16	$\cdots$	$\cdots$	$\cdots$	$\cdots$	$\cdots$
28	-0.0169	-0.0044	$\cdots$	-0.0109	$\cdots$
$\iota_{\kappa\kappa'}^{(3)}$					
1	$\cdots$	$\cdots$	$\cdots$	$\cdots$	$\cdots$
3	$\cdots$	$\cdots$	$\cdots$	$\cdots$	$\cdots$
8	0.0100	0.0028	-0.0089	-0.0077	$\cdots$
16	$\cdots$	$\cdots$	$\cdots$	$\cdots$	$\cdots$
28	-0.0122	-0.0010	$\cdots$	$\cdots$	$\cdots$
$\iota_{\kappa\kappa'}^{(4)}$					
1	$\cdots$	$\cdots$	$\cdots$	$\cdots$	$\cdots$
3	$\cdots$	$\cdots$	$\cdots$	$\cdots$	$\cdots$
8	0.0049	0.0042	-0.0101	-0.0074	$\cdots$
16	$\cdots$	$\cdots$	$\cdots$	$\cdots$	$\cdots$
28	-0.0087	0.0022	$\cdots$	-0.0086	$\cdots$
$\iota_{\kappa\kappa'}^{(5)}$					
1	$\cdots$	$\cdots$	$\cdots$	$\cdots$	$\cdots$
3	$\cdots$	$\cdots$	$\cdots$	$\cdots$	$\cdots$
8	0.0071	0.0030	-0.0081	$\cdots$	0.0015
16	$\cdots$	$\cdots$	$\cdots$	$\cdots$	$\cdots$
28	-0.0230	-0.0036	0.0003	-0.0088	$\cdots$

Table A.28 On-diagonal quartic coupling constants $\epsilon_\kappa^{(v)}$ (in eV) for the normal modes of cyclopentanone

κ	$\epsilon_\kappa^{(1)}$	$\epsilon_\kappa^{(2)}$	$\epsilon_\kappa^{(3)}$	$\epsilon_\kappa^{(4)}$	$\epsilon_\kappa^{(5)}$
1	0.0237	0.0370	0.0159	0.0223	0.0316
3	0.0049	0.0100	0.0018	0.0019	0.0072
8	0.0006	0.0005	0.0002	0.0013	0.0001
16	−0.0001	−0.0029	0.0052	0.0003	0.0019
28	$\cdots$	$\cdots$	$\cdots$	$\cdots$	$\cdots$

Table A.29 Off-diagonal linear coupling constants $\lambda_\kappa^{(v,w)}$ (in eV) for the normal modes of cyclopentanone

κ	$\lambda_\kappa^{(2,3)}$	$\lambda_\kappa^{(2,4)}$	$\lambda_\kappa^{(3,4)}$
1	0.1572	$\cdots$	0.0821
3	−0.0838	$\cdots$	−0.0156
8	$\cdots$	−0.1206	$\cdots$
16	−0.2237	$\cdots$	−0.0497
28	$\cdots$	0.0620	$\cdots$

Table A.30 Off-diagonal bilinear coupling constants $\mu_{\kappa\kappa'}^{(v,w)}$ (in eV) for the normal modes of cyclopentanone

$\mu_{\kappa\kappa'}^{(2,3)}$	1	3	8	16	28
1	$\cdots$	$\cdots$	−0.0024	$\cdots$	−0.0203
3	$\cdots$	$\cdots$	−0.0011	$\cdots$	−0.0057
8	−0.0024	−0.0011	$\cdots$	$\cdots$	$\cdots$
16	$\cdots$	$\cdots$	$\cdots$	$\cdots$	0.0063
28	−0.0203	−0.0057	$\cdots$	0.0063	$\cdots$
$\mu_{\kappa\kappa'}^{(2,4)}$					
1	−0.0052	$\cdots$	$\cdots$	0.0150	$\cdots$
3	$\cdots$	0.0042	$\cdots$	−0.0082	$\cdots$
8	$\cdots$	$\cdots$	0.0037	$\cdots$	−0.0221
16	0.0150	−0.0082	$\cdots$	−0.0081	$\cdots$
28	$\cdots$	$\cdots$	−0.0221	$\cdots$	$\cdots$
$\mu_{\kappa\kappa'}^{(3,4)}$					
1	$\cdots$	$\cdots$	0.0085	$\cdots$	$\cdots$
3	$\cdots$	$\cdots$	−0.0022	$\cdots$	0.0105
8	0.0085	−0.0022	$\cdots$	$\cdots$	$\cdots$
16	$\cdots$	$\cdots$	$\cdots$	$\cdots$	$\cdots$
28	$\cdots$	0.0105	$\cdots$	$\cdots$	$\cdots$

Table A.31 Off-diagonal linear-quadratic coupling constants $\eta_{\kappa\kappa'}^{(v,w)}$ (in eV) for the normal modes of cyclopentanone

$\eta_{\kappa\kappa'}^{(2,3)}$	1^2	3^2	8^2	16^2	28^2
1	$\cdots$	-0.0007	-0.0028	-0.0007	-0.0037
3	$\cdots$	-0.0049	0.0006	0.0029	0.0049
8	$\cdots$	$\cdots$	$\cdots$	$\cdots$	$\cdots$
16		-0.0012	0.0005	0.0118	0.0004
28	$\cdots$	$\cdots$	$\cdots$	$\cdots$	$\cdots$
$\eta_{\kappa\kappa'}^{(2,4)}$					
1	$\cdots$	$\cdots$	$\cdots$	$\cdots$	$\cdots$
3	$\cdots$	$\cdots$	$\cdots$	$\cdots$	$\cdots$
8	0.0049	0.0021	0.0071	0.0002	$\cdots$
16	$\cdots$	$\cdots$	$\cdots$	$\cdots$	$\cdots$
28	-0.0087	-0.0010	$\cdots$	$\cdots$	$\cdots$
$\eta_{\kappa\kappa'}^{(3,4)}$					
1	-0.0062	$\cdots$	-0.0003	0.0006	-0.0012
3	-0.0086	0.0041	0.0007	0.0017	-0.0009
8	$\cdots$	$\cdots$	$\cdots$	$\cdots$	$\cdots$
16	0.0007	-0.0045	-0.0003	0.0005	$\cdots$
28	$\cdots$	$\cdots$	$\cdots$	$\cdots$	$\cdots$

Appendix B
Supporting Information for Chapter 8

B.1 Conical Intersections

The geometries of the four minimum energy conical intersections (MECI) of cyclopentadiene (CPD) are depicted in Tables B.1, B.2, B.3, B.4. The values of the descriptive coordinates discussed in Chap. 8 for all the MECIs are given in Table B.5. Figure B.1 depicts the linearly interpolated (in Cartesian coordinates) and relaxed potential energy surfaces from the Franck-Condon geometry to the MECIs. The relaxed surfaces were obtained by use of a Nudged Elastic Band algorithm (NEB) [8, 9] interfaced to MOLPRO 2006.2 [10]. The NEB method is conventionally used for finding the minimum energy path between a pair of local minima or stable states, however, inhere it is used where at least one of the endpoints is not such a configuration. This use of NEB can lead to ambiguity for the path around the endpoint. However, the method is only invoked to demonstrate that barriers present on interpolated potential energy surfaces might not be present if the surfaces are relaxed, and as such whether the NEB method converges to the actual minimum energy path is not of importance here. Figure B.2 shows the potential energy surfaces of the MECIs in the branching space.

T. S. Kuhlman, *The Non-Ergodic Nature of Internal Conversion*, Springer Theses,
DOI: 10.1007/978-3-319-00386-3, © Springer International Publishing Switzerland 2013

Table B.1 Geometry of the *eth*1 MECI (in Å)

Atom	x	y	z
C	−0.237305	−0.045771	1.333271
C	0.138831	1.076571	0.356015
C	0.163163	−1.083575	0.351376
C	−0.489140	−0.749858	−0.974490
C	−0.095690	0.601737	−0.976102
H	−1.313034	−0.065797	1.536982
H	0.339337	−0.036818	2.255284
H	0.756636	1.917868	0.634213
H	1.201292	−1.430628	0.375283
H	−0.508682	−1.398627	−1.836017
H	0.044593	1.214900	−1.860533

Table B.2 Geometry of the *eth*2 MECI (in Å)

Atom	x	y	z
C	−0.236206	−0.032080	1.291252
C	0.215824	1.011314	0.362281
C	0.116410	−1.325582	0.406327
C	−0.234085	−0.755477	−0.972845
C	−0.048608	0.632833	−1.006990
H	−1.323021	−0.020961	1.404294
H	0.240820	−0.010103	2.270001
H	0.818225	1.873797	0.644010
H	1.219464	−1.336168	0.427286
H	−0.622274	−1.355418	−1.790937
H	−0.146550	1.317845	−1.839398

Table B.3 Geometry of the *dis* MECI (in Å)

Atom	x	y	z
C	0.257136	−0.014429	1.318018
C	−0.039247	1.209763	0.385995
C	−0.197883	−1.031573	0.362890
C	0.336413	−0.706479	−0.999024
C	0.147293	0.655824	−0.995520
H	−0.282847	0.008378	2.261221
H	1.335113	−0.086087	1.491612
H	−1.017542	1.652186	0.577767
H	−1.121552	−1.596069	0.499694
H	0.425101	−1.383974	−1.832478
H	0.158015	1.292460	−1.874892

Table B.4 Geometry of the S_2S_1 MECI (in Å)

Atom	x	y	z
C	−0.222312	0.000055	1.295210
C	0.173705	1.094371	0.345488
C	0.173641	−1.094866	0.345542
C	−0.179011	−0.698783	−1.007255
C	−0.178394	0.698547	−1.007401
H	−1.302930	−0.001162	1.504808
H	0.331448	0.001044	2.231690
H	0.908564	1.853016	0.594770
H	0.910036	−1.852080	0.594095
H	−0.309099	−1.362005	−1.850634
H	−0.305647	1.361864	−1.851033

Table B.5 Values of the bond alternation, backbone torsion, C=C twist and CH_2 bend for the four MECIs of CPD

Coordinate	*eth*1	*eth*2	*dis*	S_2S_1
Bond alternation/Å	1.54	1.58	1.62	1.62
Backbone torsion/deg	24.5	8.5	18.3	0.0
C=C twist/deg	50.5	58.9	43.4	31.7
CH_2 bend/deg	40.1	36.9	37.0	37.2

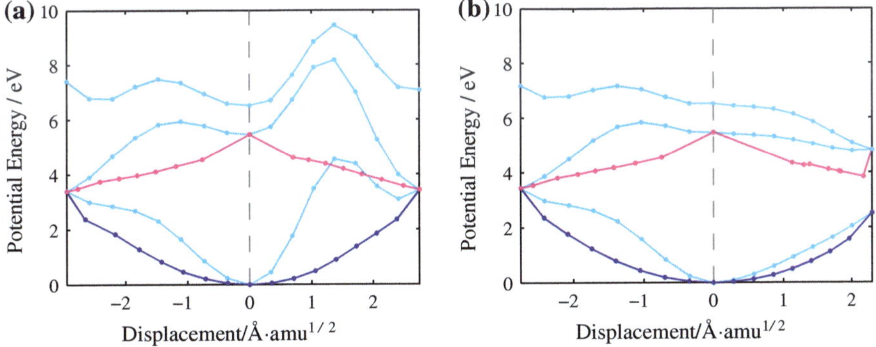

Fig. B.1 Potential energy surfaces for CPD connecting the Franck-Condon geometry with the four MECIs as a function of mass-weighted displacement. Linearly interpolated (in Cartesian coordinates) potential energy surfaces (*cyan*), along with the S_0 (*purple*) and S_1 (*magenta*) potential energy surfaces resulting from relaxation on the S_1 state (T. S. Kuhlman et al., Faraday Discuss. **157**, 193–212 (2012)—Adapted by permission of The Royal Society of Chemistry). (**a**) *eth*1 MECI ← FC → *eth*2 MECI. (**b**) *dis* MECI ← FC → S_2S_1 MECI

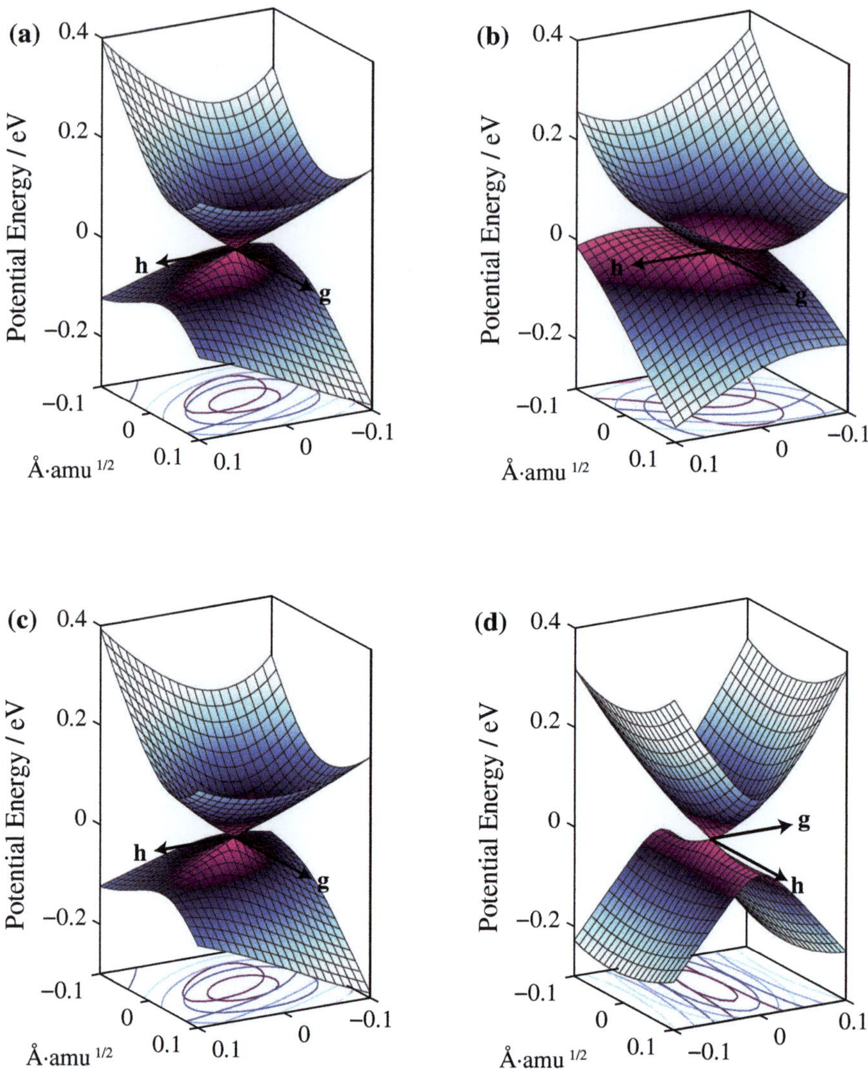

Fig. B.2 MECIs of CPD in the branching space of the scaled derivative coupling and gradient difference vectors. All MECI exhibit a peaked geometry (T. S. Kuhlman et al., Faraday Discuss. **157**, 193–212 (2012)—Adapted by permission of The Royal Society of Chemistry). (**a**) *eth*1 MECI. (**b**) *eth*2 MECI. (**c**) *dis* MECI. (**d**) S_2S_1 MECI

References

1. K. Kaufmann, W. Baumeister, M. Jungen, J. Phys. B **22**, 2223–2240 (1989)
2. B.O. Roos, K. Andersson, M.P. Fülscher, P.-Å. Malmqvist, L. Serrano-Andrés, P. Pierloot, M. Merchán, Adv. Chem. Phys. **93**, 219–331 (1996)
3. M.J. Frisch et al., *Gaussian 03, Revision E.01* (Gaussian, Inc., Wallingford, 2004)
4. M.J. Frisch et al., *Gaussian 09, Revision B.01* (Gaussian, Inc., Wallingford, 2009)
5. J.F. Stanton et al., Cfour, a quantum chemical program package. For the current version, see http://www.cfour.de
6. L. Gagliardi, R. Lindh, G. Karlström, J. Chem. Phys. **121**, 4494–4500 (2004)
7. G. Karlström, R. Lindh, P.-Å. Malmqvist, B.O. Roos, U. Ryde, V. Veryazov, P.-O. Widmark, M. Cossi, B. Schimmelpfennig, P. Neogrady, L. Seijo, Comput. Mater. Sci. **28**, 222–239 (2003)
8. H. Jónsson, G. Mills, K.W. Jacobsen, Nudged Elastic Band Method for Finding Minimum Energy Paths of Transitions, in *Classical and Quantum Dynamics in Condensed Phase Simulations*, ed. by B.J. Berne, G. Ciccotti, D.F. Coker (World Scientific, Singapore, 1998), pp. 385–404
9. D. Sheppard, R. Terrell, G. Henkelman, J. Chem. Phys. **128**, 134106 (2008)
10. H.-J. Werner, P.J. Knowles, R. Lindh, F.R. Manby, M. Schütz, others, Molpro, version 2006.2, a package of ab initio programs, see http://www.molpro.net

MIX
Papier aus verantwortungsvollen Quellen
Paper from responsible sources
FSC® C105338

If you have any concerns about our products,
you can contact us on
ProductSafety@springernature.com

In case Publisher is established outside the EU,
the EU authorized representative is:
Springer Nature Customer Service Center GmbH
Europaplatz 3, 69115 Heidelberg, Germany

Printed by Libri Plureos GmbH
in Hamburg, Germany